MATHEMATIQUES
&
APPLICATIONS

Directeurs de la collection:
X. Guyon et J.-M. Thomas

37

Springer
Paris
Berlin
Heidelberg
New York
Barcelone
Hong Kong
Londres
Milan
Tokyo

Instructions aux auteurs:

Les textes ou projets peuvent être soumis directement à l'un des membres du comité de lecture avec copie à X. GUYON ou J.-M. THOMAS. Les manuscrits devront être remis à l'Éditeur *in fine* prêts à être reproduits par procédé photographique.

Sylvain Sorin

A First Course on Zero-Sum Repeated Games

Springer

Sylvain Sorin
Laboratoire d'Econométrie
Ecole Polytechnique
1, rue Descartes
75005 Paris, France

and

Equipe Combinatoire, UFR 921
Université Pierre et Marie Curie – Paris 6
175, rue du Chevaleret
75013 Paris, France

sorin@poly.polytechnique.fr

Mathematics Subject Classification 2000: 91A05l, 91A10, 91A15, 91A20

ISBN 3-540-43028-8 Springer-Verlag Berlin Heidelberg New York

Springer-Verlag Berlin Heidelberg New York
est membre du groupe BertelsmannSpringer Science+Business Media GmbH.

http://www.springer.de
Imprimé en Allemagne

Imprimé sur papier non acide SPIN: 10762997 41/3142/So - 5 4 3 2 1 0 -

Foreword

This course deals with two-person zero-sum repeated games. These are multi-move games played in stages . Alternatively, they may be viewed as discrete-time stochastic processes controlled by both players with partial observability. The main assumption is stationarity: the basic structure does not change over time.

Among the games in this class, stochastic games have been studied first. Introduced by Shapley in 1953, they are described as follows. A collection of states Ω is given. At stage n, the state ω_n, known by all players, and the actions of the players determine both a real payoff and a random transition to a new state ω_{n+1}.

Another important family corresponds to incomplete information games. They were initialy defined and analyzed by Aumann and Maschler in 1966. Again a collection of states is given but in addition it is endowed with some probability. In this framework, the state initialy chosen remains constant and the players receive only partial information about it.

Hence, in the first case the state is changing but known, while it is constant and unknown in the second. Stochastic games are concerned with the joint control of a random state while incomplete information games deal with the issue of extracting or revealing information.

In addition, in both classes different kinds of repeated games have been studied. The main interest in this course concerns asymptotic properties, but there are several ways of modeling long term games. Three approaches will be considered: limits of long finite games, infinite games with vanishing discount factors, and uniform properties in infinite undiscounted games.

In the first two cases, the minmax theorem leads to a basic recursive formula already established by Shapley for stochastic games.

In the third approach the analysis is based on an explicit construction of robust strategies.

The main purposes of these lectures are:

- to present the basic results, especially for games with incomplete information and stochastic games;
- to show the large variety of tools and approaches available (different proofs are provided for several results); and
- to give insights and explanations concerning extensions, recent develop-

ments and directions of research.
The contents of the course are briefly described as follows.
Chapter 1 introduces two-person zero-sum games and studies a collection of examples illustrating some of the topics to be studied: use of information, asymptotic properties.
Chapter 2 emphasizes the connection between convexity and incomplete information games (ignoring the "transmission of information" aspects that will occur with the repetition of the game).
Chapter 3 presents the study of repeated games with incomplete information on one side. The asymptotic approach is introduced first. The stronger existence of uniform value is then obtained.
Chapter 4 is devoted to the case of lack of information on both sides. The minmax and maxmin of the infinite game are identified, which usually are different. The existence of the asymptotic value is then shown.
Chapter 5 contains a short presentation of the basic results for zero-sum stochastic games, especially the existence proof of uniform value in the finite case.
The aim of Chapter 6 is to introduce several models and results illustrating the relation between incomplete information and stochastic games.
In addition, there are four appendices.
Appendix A recalls basic facts on two-person zero-sum games and gives a selection of minmax theorems.
Appendix B deals briefly with approachability theory, due to Blackwell (1956) which is fundamental in the study of incomplete information games.
Appendix C presents more recent advances related to the operator approach that are used in the development of several topics.
Appendix D is devoted to a specific simple proof of Kuhn's theorem (1953) for the class of games under consideration.

These lectures are an introduction to the field and deal with the simplest models: for example standard signalling is always assumed (the past moves are known) and the general case of payoff functions defined on plays is not treated, see Maitra and Sudderth (1996). Some extensions are discussed in the notes after each chapter.
Let us also underline the fact that only zero-sum games are covered. There are several reasons to justify this choice.
First, the study of this case is a necessary step before considering n-person games. It focuses attention on the antagonistic aspects and avoids taking into account incentives for cooperation, so is easier to interpret.
Second, the analysis of situations with opposite interests is needed to determine the individually rational levels on which are based the threats that sustain equilibria in non zero-sum repeated games.
Third, the non zero-sum case presents conceptual developments and specific problems that are best avoided in an introduction, including:
- selection and/or extension of equilibria: Kohlberg and Mertens (1986), Au-

mann (1974);
- extensive form correlation, communication and protocols: Forges (1986, 1990):
- robustness of equilibria in long games, asymptotic versus infinite approach: Sorin (1986).

Fourth, the nature of the mathematical tools used for the study of the non-zero sum case is quite different, in the incomplete information case :
- bi-martingale and biconvexity: Hart (1985), Aumann and Hart (1986); and
- algebraic topology: Simon, Spiez and Torunczyk (1995), and for stochastic games
- approximations of fixed points: Mertens and Parthasarathy (1987); and
- control of perturbations: Vieille (2000a, 2000b, 2000c), Solan (1999).

This course is in the spirit of introductory notes, dealing with incomplete information games in the simplest case (Sorin, 1979) where the hypotheses of standard signalling and independent case are assumed. In particular much of the material produced by Aumann, Maschler and Stearns in the ACDA 1966-1968 reports (Aumann and Maschler 1995) is not treated here. A fortiori, only a part of the content of the survey by Zamir (1992) is covered.
On the other hand stochastic games are presented and the connection between the two domains is emphasized, which in some cases extends to the non zero-sum case, Neyman and Sorin (1998). I tried to illustrate the basic rôle of the recursive structure and the corresponding operator approach. Recent advances such as the dual game and connections with differential games are also introduced.
My aim is to describe a large accessible panorama to the reader interested in discovering the area. I hope he will like it and be encouraged to enjoy the lecture of the more advanced presentation by Mertens, Sorin and Zamir (1994) (henceforth referred to as MSZ).

Thanks

Some parts of this manuscript correspond to graduate courses given in several doctoral programs during the last 20 years: Université Paris 9-Dauphine, ENSAE, Université Louis Pasteur-Strasbourg 1, Ecole Normale Supérieure, Université Pierre et Marie Curie-Paris 6, Ecole Polytechnique, Université Paris 10-Nanterre, CORE (Louvain La Neuve), SUNY at Stony Brook, IMPA (Rio) and Universidad de Chile (Santiago) as well as crash courses at LSE (London), IMW (Bielefeld), IMSSS (Stanford), Haiffa, Vigo, La Habana. In each case, I greatly benefited from the comments of the audience.

I very much appreciated the remarks and suggestions of R. Laraki, E. Lehrer, A. Neyman, D. Rosenberg, E. Solan, W. Sudderth and N. Vieille.
Finally, my deep debt towards my friends and colleagues J.-F. Mertens and S. Zamir is clear, and recognized with pleasure.

Part of the writing was done while I was visiting IMPA and the Universidad de Chile (FONDAP de Matematicas Aplicadas). The support of these institutions is gratefully acknowledged.

References

Aumann R.J. (1974) Subjectivity and correlation in randomized strategies, *Journal of Mathematical Economics*, **1**, 67-95.

Aumann R.J. and S. Hart (1986) Bi-convexity and bi-martingales, *Israel Journal of Mathematics*, **54**, 159-180.

Aumann R.J. and M. Maschler (1995) *Repeated Games with Incomplete Information*, M.I.T. Press (with the collaboration of R. Stearns).

Blackwell D. (1956) An analog of the minmax theorem for vector payoffs, *Pacific Journal of Mathematics*, **6**, 1-8.

Forges F. (1986) An approach to communication equilibria, *Econometrica*, **54**, 1375-1385.

Forges F. (1990) Universal mechanisms, *Econometrica*, **58**, 1341-1364.

Hart S. (1985) Non zero-sum two-person repeated games with incomplete information, *Mathematics of Operations Research*, **10**, 117-151.

Kohlberg E. and J.-F. Mertens (1986) On the strategic stability of equilibria, *Econometrica*, **54**, 1003-1037.

Kuhn H. W. (1953) Extensive games and the problem of information, in *Contibutions to the Theory of Games, II*, H.W. Kuhn and A.W. Tucker (eds.), Annals of Mathematical Studies, 24, Princeton University Press, 193-216.

Maitra A. and W. Sudderth (1996) *Discrete Gambling and Stochastic Games*, Springer.

Mertens J.-F.and T. Parthasarathy (1987) Existence and characterization of Nash equilibria for discounted stochastic games, CORE D.P. 8750.

Mertens J.-F., S. Sorin and S. Zamir (1994) *Repeated Games*, CORE D.P. 9420-21-22.

Neyman A. and S. Sorin (1998) Equilibria in repeated games of incomplete information: the general symmetric case, *International Journal of Game Theory*, **27**, 201-210.

Shapley L. S. (1953) Stochastic games, *Proceedings of the National Academy of Sciences of the U.S.A*, **39**, 1095-1100.

Simon R.S., S. Spiez and H. Torunczyk (1995) The existence of an equilibrium in games of incomplete information on one side and a theorem of Borsuk-Ulam type, *Israel Journal of Mathematics*, **92**, 1-21.

Solan E. (1999) Three player absorbing games, *Mathematics of Operations Research*, **24**, 669-698.

Sorin S. (1979) An introduction to two-person zero-sum repeated games with incomplete information, *Cahiers du Groupe de Mathématiques Economiques*, **1** (English version, TR 312, IMSS-Economics, Stanford University, 1980).

Sorin S. (1986) Asymptotic properties of a non zero-sum stochastic game, *International Journal of Game Theory*, **15**, 101-107.

Vieille N. (2000a) Two player stochastic games I: A reduction, *Israel Journal of Mathematics*, **119**, 55-91.

Vieille N. (2000b) Two player stochastic games II: The case of recursive games, *Israel Journal of Mathematics*, **119**, 93-126.

Vieille N. (2000c) Large deviations and stochastic games, *Israel Journal of Mathematics*, **119**, 127-142.

Zamir S. (1992) Repeated games of incomplete information: zero-sum in *Handbook of Game Theory, I*, R.J. Aumann and S. Hart (eds.), North Holland, 109-154.

Notations

For x in $\mathbb{R}^n$, $x \geq 0$ means $x_i \geq 0, \forall i$, $x \gg 0$ corresponds to $x_i > 0, \forall i$ and finally $x > 0$ stands for $x \neq 0$ and $x \geq 0$.
$\|x\|_1 = \sum |x_i|$
$\|x\|_\infty = \sup_i |x_i|$
Given $(x_1, x_2, ..., x_n) \in \mathbb{R}^K$, $\overline{x}_n = \dfrac{1}{n} \sum_{i=1}^{n} x_i$
$sAt = \sum_{ij} s_i A_{ij} t_j$, where A is a $m \times n$ matrix and $x \in \mathbb{R}^m$, $y \in \mathbb{R}^n$
$\langle , \rangle$: scalar product

$\# A$ = cardinal of the set A
$\overline{A}$: closure of A
$intA$: interior of the set A
$\mathbf{1}_A$: indicator of the set A
co: convex hull

$\Delta(K)$ for K a finite set, denotes the simplex of probabilities on K.
More generally, $\Delta(X)$ for X a topological space, denotes the set of Borel regular probabilities endowed with the weak * topology.
$\Delta_f(X)$ is the subset of probabilities with finite support.
For μ measure on a countable set M $[\mu] = \max_{m \in M} |\mu(m)|$
$[[\mu]] = \|\mu - \theta * \mu\|$

I, J: moves sets
$S = \Delta(I), T = \Delta(J)$: mixed move sets
$X = S^K$, $Y = T^L$

$\mathtt{val}$: value operator
$\mathtt{val}_{X \times Y} f = \sup_X \inf_Y f(x,y) = \inf_Y \sup_X f(x,y)$
$\mathtt{Cav}\ f$: concavification of the function f
$\mathtt{Vex}\ f$: convexification of the function f
∇f: gradient of the function f
$\mathtt{epi}(f)$: epigraph of the function f
$\mathtt{Dom}(f)$: domain of the function f
end of proof: ■

Table of Contents

1 Introduction and examples

The purpose of this chapter is to present several examples illustrating some of the problems under consideration. A preliminary Section recalls briefly basic facts on two-person zero sum games. In Section 2 a first collection of examples deals with the use of information. Section 3 is concerned with the notion of value in long games. Miscellaneous observations are gathered in Section 4. Some basic references in game theory are given in Section 5.

1.1 Two-person zero-sum matrix games

This section is an elementary introduction to two-person zero sum matrix games. A more general presentation of two-person zero-sum games and of related minmax theorems with proofs can be found in Appendix A.

A **two-person zero-sum matrix game** is defined by a real matrix $A = ((A_{ij})), i{\in}I, j{\in}J$. The set I of lines (resp. J of columns) is the set of **strategies** of Player 1 who is the maximizer (resp. Player 2 who is the minimizer). A choice i by Player 1 and j by Player 2 determines a **payoff** A_{ij}. i being given, Player 1 will get at least $\min_{j\in J} A_{ij}$ hence by an optimal choice of i he can achieve

$$\underline{w} = \max_{i\in I} \min_{j\in J} A_{ij}$$

which is called the **maxmin** (in pure strategies) of the game A. Similarly Player can obtain a payoff bounded by

$$\overline{w} = \min_{j\in J} \max_{i\in I} A_{ij}.$$

which corresponds to the **minmax**. It is clear that the inequality $\underline{w} \leq \overline{w}$ always holds but it can be strict. In the game below

	L	R
T	1	0
B	0	1

the strategy sets are $I = \{T, B\}$ (for Top and Bottom), $J = \{L, R\}$(Left, Right). Given any choice of Player 1, Player 2 can obtain 0 and $\underline{w} = 0$. On the other hand, given any choice of Player 2, Player 1 can get 1 and $\overline{w} = 1$. In the next game

	L	R
T	0	1
B	-1	x

$\overline{w} = \underline{w} = 0$, for all x in $I\!R$, hence a **value** exists.
In the first game the knowledge in advance of the move of the opponent (the order in which they play) is relevant. In the second one, it is not: Player 1 can announce his choice T and be sure of achieving at least 0. Similarly for Player 2, playing L insures a payoff at most 0.
Let us now allow the choice of a strategy by a player to be random. The set of **mixed strategies** of Player 1 is the simplex on I: $S = \Delta(I) = \{s \in I\!R^I : \sum_{i \in I} s_i = 1, s_i \geq 0, i \in I\}$. An element $s \in S$ is a probability on the set I of moves. Similarly mixed moves of Player 2 form the set $T = \Delta(J) = \{t \in I\!R^J : \sum_{j \in J} t_j = 1, t_j \geq 0, j \in J\}$. The payoff corresponding to s and t is the bilinear extension (or equivalently the expectation) $sAt = \sum_{i \in I, j \in J} s_i A_{ij} t_j$. The triple (S, T, A) is the **mixed extension** of the game A. It is thus natural to consider the new maxmin and minmax, namely: $\underline{v} = \max_{s \in S} \min_{t \in T} sAt$ and $\overline{v} = \min_{t \in T} \max_{s \in S} sAt$. By linearity $\min_{t \in T} sAt = \min_{j \in J} sA_j$ (J is the set of extreme points of the convex set T, where the move j is identified with the Dirac mass at j) hence $\underline{v} \geq \underline{w}$. The use of mixed strategies thus reduces the "duality gap":

$$\overline{v} - \underline{v} \leq \overline{w} - \underline{w}.$$

The basic minmax theorem due to von Neumann proves that in the new game the equality maxmin = minmax is satisfied. Hence the game has a **value** v, with $v = \underline{v} = \overline{v}$.
Again the interpretation in terms of order of play (or information) is the following: Player 1 can announce his strategy (namely x) in advance, he will still guarantee the value. Note that if x is understood as the distribution of the actual move i, this means that the law of the random variable i is known, but obviously not its realization.
The minmax's theorem (Theorem A.5) can be stated in terms of the **value operator** `val` defined on real matrices by $\mathtt{val}A = v$:
Given a real $I \times J$ matrix A, there exists a real number $\mathtt{val}A$ and elements $\overline{s}$ in $S = \Delta(I)$ and $\overline{t}$ in $T = \Delta(J)$ satisfying

$$\overline{s}At \geq \mathtt{val}A, \qquad \forall t \varepsilon T$$

$$sA\overline{t} \leq \mathtt{val}A, \qquad \forall s \in S.$$

Such strategies $\overline{s}$ of Player 1 and $\overline{t}$ of Player 2 are called **optimal**.
In the above first example the value is 1/2 and optimal strategies for the players are $(1/2, 1/2)$. In the next one

	L	R
T	2	-1
B	0	1

the value is $1/2$ and optimal strategies are $(1/4, 3/4)$ and $(1/2, 1/2)$ respectively.

More generally a **two-person zero-sum game** is a triple $(g; X, Y)$ where g is a real function defined on a product space $X \times Y$. X (resp. Y) is the **strategy set** of the maximizer, Player 1 (resp. of the minimizer, Player 2) and g is the **payoff function.** One defines the maxmin by by

$$\underline{v}(g) = \sup_X \inf_Y g(x, y)$$

the minmax by

$$\overline{v}(g) = \inf_Y \sup_X g(x, y)$$

and the inequality $\underline{v} \leq \overline{v}$ always holds.
Given $\varepsilon \geq 0$, a strategy x of Player 1 is **ε-optimal** if:

$$g(x, y) \geq \underline{v} - \varepsilon, \quad \forall y \in Y.$$

Similarly a strategy y of Player 2 is **ε-optimal** if:

$$g(x, y) \leq \overline{v} + \varepsilon, \quad \forall x \in X.$$

Such strategies always exist for $\varepsilon > 0$. If they exist for $\varepsilon = 0$, they are called **optimal.**
A **minmax theorem** gives conditions on $(g; X, Y)$ for the game to have a value, namely for the equality $\underline{v} = \overline{v}$ to be satisfied. This quantity is then written $v = \mathtt{val}_{X \times Y} g$.
Under suitable measurability and integrability requirements one defines a **mixed extension** of (g, X, Y) as a game $(\gamma, \Sigma, \mathcal{T})$ where Σ and $\mathcal{T}$ are convex subsets of the set of probabilities on X and Y, and the payoff γ is the bilinear extension:

$$\gamma(\sigma, \tau) = \int_{X \times Y} g(x, y) \sigma(dx) \otimes \tau(dy).$$

Then the minmax and maxmin are defined accordingly.

The repetition of a zero-sum game has no specific interest. If a player uses an optimal strategy i.i.d. at each stage he can guarantee the value at that stage. The (normalized) value of any kind of repetition is the value of the one stage game.
Hence we will be interested in situations where the basic structure is stationary but some parameter (information or state) will vary with the repetition.

1.2 On the use of information

In the three examples below, two players are playing repeatedly a zero-sum matrix game. However this game is chosen at random (once for all according

to some probability P) and only Player 1 (the maximizer and row player) knows this choice. Both players know P. In addition after each stage of the game both players are told the previous moves, but not the payoff that could reveal some information on the game selected. This description of the rules is known by both.
We focus here on the behavior of the informed player: how should he take advantage of his private information, especially in the long run?

The first case is given by the payoff matrices below, which are chosen with initial probability $(1/2, 1/2)$:

	L	R
T	1	0
B	0	0

G^1

	L	R
T	0	0
B	0	1

G^2

The one shot game is thus described by the next matrix:

	L	R
TT	1/2	0
TB	1/2	1/2
BT	0	0
BB	0	1/2

Indeed a strategy for Player 1 in the game is a choice of a move as a function of his information; hence TB stands for the strategy consisting in playing T in game G^1 and B in game G^2 and so on ....
The optimal strategy of player 1 is thus to play TB and the value of the one stage game is 1/2. However the use at stage one, of this strategy in long games is not so interesting. In fact such a behavior is **completely revealing** (CR) in the sense that the knowledge of the strategy and the observation of the move allows to identify the true game being played: G^1 if T is played and G^2 otherwise. From then on, Player 2 can force a payoff $0 = \mathtt{val}\ G^1 = \mathtt{val}\ G^2$ at each stage. The asymptotic average payoff is then 0.
On the other hand Player 1 may disregard entirely his information, in which case each stage of the game amounts to playing according to the following matrix:

1/2	0
0	1/2

which corresponds to the expected payoff. In this game, with value 1/4, Player 1 can guarantee, by playing i.i.d. an optimal strategy $(1/2, 1/2)$, a stage payoff of 1/4, hence the same amount on the average.

Thus in the current case a **non revealing** (NR) strategy is better than a completely revealing one.

Assume now the following payoffs:

−1	0
0	0

G^1

0	0
0	−1

G^2

A similar analysis shows that a CR strategy will guarantee asymptotically 0 to Player 1, which is obviously the best he can get. By playing an optimal NR strategy he would only achieve the value of the average game:

−1/2	0
0	−1/2

which is $-1/4$.

Consider finally the next case where Player 2 has 3 moves, $J = \{L, M, R\}$:

	L	M	R
T	4	0	2
B	4	0	−2

G^1

	L	M	R
T	0	4	−2
B	0	4	2

G^2

Since $\mathtt{val}\ G^1 = \mathtt{val}\ G^2 = 0$, a CR strategy will guarantee asymptotically 0. On the other hand the average game is given by:

	L	M	R
T	2	2	0
B	2	2	0

with value again 0, which is thus the best amount obtained by a NR strategy. We now exhibit a strategy of Player 1 that will do better. Consider two auxiliary lotteries on the letters (t, b): P^1 where the probability is $(3/4, 1/4)$ and P^2 where it is $(1/4, 3/4)$. Player 1 uses the following **type-dependent** procedure: if the game is G^k $(k = 1, 2)$, perform the lottery P^k. If the outcome is t play T for ever and B for ever, otherwise.

Assume even that the above behavior of Player 1 is known by Player 2. After a move T of Player 1 at stage 1, Player 2 anticipates Player 1 playing T in the future and his new posterior distribution on the state is $(3/4, 1/4)$. Explicitly

$$Prob(k = 1|T) = \frac{Prob(k = 1, T)}{Prob(T)} = \frac{Prob(k = 1)Prob(T|k = 1)}{Prob(T)}$$

$$= \frac{(1/2)(3/4)}{(1/2)(3/4) + (1/2)(1/4)} = 3/4.$$

He is thus facing at each stage the following expected payoff:

	L	M	R
T	3	1	1

Similarly if the first move is B, Player 2 is computing the expected payoff:

	L	M	R
B	1	3	1

so that in both situations Player 1 can achieve a stage payoff of 1. The use of a partially revealing strategy is thus better for him.

More generally along a play of the game the strategy of the informed Player will generate a **martingale of posterior distributions** and this will be one of the basic tools in the analysis of the repeated game. The NR and CR strategies correspond to the two extreme cases of a constant martingale or of a jump at stage one to extreme points. In the figure below p is the initial probability of G^1.

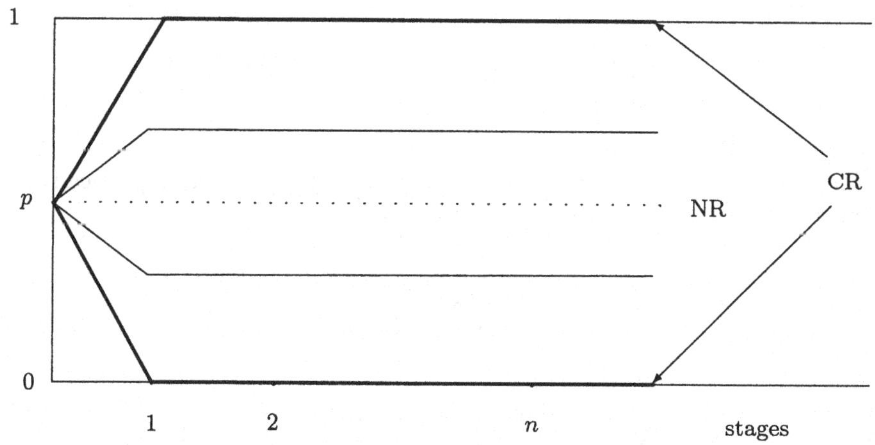

Fig. 1.1. The martingale of posterior distribution

1.3 On the notion of value in long games

In this section we introduce and compare two possible evaluations of the stream of stage payoffs in long games.

One way is to consider a sequence of n stage repeated games, each having a value v_n and to study the asymptotic behavior of the sequence $\{v_n\}$. If it

converges it leads to the asymptotic value.
An other approach is to compute for each of the strategies of Player 1 the amount it guarantees in long games and to take as maxmin the supremum of such quantities. A dual definition holds for Player 2 and for the minmax. If these quantities coincide they will define the uniform value, v_∞. (Precise definitions will be provided in Chapter 3).

Consider the game where two matrices G^1 and G^2 are given. $k = 1, 2$ is chosen equally at random, none of the players being informed. Both players play then repeatedly the matrix game. The payoff is computed according to G^k. In addition, there is a public signal at each stage determined by the signalling matrix H^k with the following data:

$$G^1 = \begin{array}{|c|c|}\hline 0 & 8 \\ \hline 0 & 8 \\ \hline\end{array} \qquad H^1 = \begin{array}{|c|c|}\hline \alpha & \alpha \\ \hline b & c \\ \hline\end{array}$$

$$G^2 = \begin{array}{|c|c|}\hline 8 & 0 \\ \hline 8 & 0 \\ \hline\end{array} \qquad H^2 = \begin{array}{|c|c|}\hline \alpha & \alpha \\ \hline d & c \\ \hline\end{array}$$

For example, if $k = 2$ and (B, L) is played the signal to both players is d. Note that the signal α informs Player 2 that Player 1 played T, while Player 1 does not know the move of Player 2. c reveals the pair of moves while b or d reveal in addition the state k.
Player 1 can achieve an expected payoff of 4 by playing Top at each stage (k is not revealed).
Player 2 can as well guarantee 4 by playing $(1/2, 1/2)$ i.i.d.. Hence max min $=$ min max so that v_∞ exists and equals 4.

Consider now the case of three states with the following features (and the same rules): $K = \{1, 2, 3\}$, p in $\Delta(K)$ is the initial probability, G^k the payoff matrix and H^k the signalling matrix.

$$p^1 = 1/4, \quad G^1 = \begin{array}{|c|c|}\hline 0 & 8 \\ \hline 0 & 8 \\ \hline\end{array} \qquad H^1 = \begin{array}{|c|c|}\hline \alpha & \alpha \\ \hline b & c \\ \hline\end{array}$$

$$p^2 = 1/4, \quad G^2 = \begin{array}{|c|c|}\hline 8 & 0 \\ \hline 8 & 0 \\ \hline\end{array} \qquad H^2 = \begin{array}{|c|c|}\hline \alpha & \alpha \\ \hline d & c \\ \hline\end{array}$$

$$p^3 = 1/2, \quad G^3 = \begin{array}{|c|c|}\hline 0 & -4 \\ \hline 0 & 0 \\ \hline\end{array} \qquad H^3 = \begin{array}{|c|c|}\hline \alpha & \alpha \\ \hline e & f \\ \hline\end{array}$$

By playing $(1/2, 1/2)$ i.i.d. Player 2 obtains up to θ, defined as the first stage where Player 1 plays Bottom, an expected payoff of 1: namely 4 in G^1 and

G^2 and -2 in G^3. If at stage θ, a letter different from c occurs, the game G^k is revealed: explicitly G^1 if b, G^2 if d and G^3 if e or f. Since both players know the true game, one can assume that they play optimally and the payoff from then on is $0 = \texttt{val}\ G^k, \forall k$. Finally the probability that the letter c will occur at stage θ is 1/4 and the subsequent payoff is at most 4, since the game given c is the one previously studied: both players know that k is either 1 or 2 and the conditional probability is $(1/2, 1/2)$. Hence Player 2 can guarantee 1.

If Player 1 knows the strategy of Player 2, he can achieve 1 as follows:
- play Top as long as the probability of Left is at least 1/2.
- play Bottom as soon as it goes below 1/2 and then optimally in the revealed game, depending on the signal.

The expected payoff as long as Top is played is at least 1 (since it is at least -2 in G^3). When Bottom is played the probability of c is at least 1/4, the other letters are revealing and the previous argument applies giving an expected payoff at least 1.

The above computations show that:

$$\lim_{n \to \infty} v_n = 1 = \min\max .$$

On the other hand one will prove that:

$$\max\min = 0.$$

In fact, given σ a strategy of Player 1, let π be the first stage where Player 1, while facing a sequence of L of Player 2, plays Bottom; and let ρ be the probability of the event $\{\pi < \infty\}$. Define a strategy τ of Player 2 by: play Left until stage π (included) and then optimal in the revealed game. If $\rho = 1$, the game will eventually be revealed in finite time (since b, d, e, are revealing) and from stage π on, the payoff is 0. If $\rho < 1$, given $\varepsilon \geq 0$, there exists N such that:

$$P_{\sigma,\tau}(\pi < \infty | \pi \geq N) \leq \varepsilon.$$

The point is that a change of strategy by Player 2, from L to R after stage N, will remain undetected by Player 1 until stage π, since the signals will be the same; hence the same upper bound applies under the alternative strategy τ' of Player 2: play L until N then always R. Thus if π is less than N the asymptotic payoff is 0, and if π is larger than N the payoff after N is with probability at least $(1-\varepsilon)$, $(1/4)\times 8 + (1/4)\times 0 + (1/2)\times(-4) = 0$. Hence for n large enough, the average expected payoff up to stage n is less than $0(\varepsilon)$.

It follows that the two notions of value actually differ: in the asymptotic approach ($\lim v_n$) the strategy can be a f6nction of the length of the game, which is not the case in the uniform approach. Explicitely when dealing with asymptotic properties we are looking for an "asymptotic game" that would approximate any long interaction with publicly known duration.

On the other hand uniform properties are required to hold in any sufficiently long game, independently of its real (unknown) length and correspond to robustness of the strategic behavior.

Note also that the previous game is asymptotically equivalent to the following "absorbing game" with no signals (see Chapter 6, Section 6.4):

2	0
0*	2*

α	α
.	.

where a * denotes an absorbing payoff. Here if Player 2 announces his strategy he cannot get less than 1 and if Player 1 announces his own he cannot obtain more than 0. This game also provides an exemple of connection between incomplete information and stochastic games, see Chapter 6.

1.4 Miscellaneous

Markov Decision Processes (MDP) or dynamic programming correspond to one player games and some phenomena are easier to illustrate in this framework, see also Chapter 5, Section 5.7. The next examples deal with comparison of evaluations an exitence of optimal strategies.

1.4.1 Evaluations in Markov decision processes

Consider the following deterministic MDP. There is a state space Ω and for each state ω a set of available moves $A(\omega)$. A couple of state and move (ω, a) determines a current payoff and a new state ω'. The description is as follows: At each state or node of the form $(n, 0)$ one can choose Right leading to $(n+1, 0)$ or Top leading to $(n, 1)$. At each node (n, m), $m \geq 1$ the transition is deterministic to $(n, m+1)$. The payoff is 0 except at nodes (n, m), with $1 \leq m \leq n$ where it is 1. The initial state is $(1, 0)$. Denote by g_m the gain at stage m and consider the two following evaluations of the sequence $\{g_m\}$.
The n-stage game G_n is the game with payoff $\overline{g}_n = \frac{1}{n}\sum_{m=1}^{n} g_m$, the Cesaro average of the payoff during the first n stages and its value is v_n.
In the λ-discounted game $G_\lambda, \lambda \in (0, 1]$, the payoff is $g_\lambda = \sum_{m=1}^{\infty} \lambda(1 - \lambda)^{m-1} g_m$ and the value is v_λ.
Since all feasible paths are of the form "Right until stage θ then Top" ($\theta \in \mathbb{N} \cup \{+\infty\}$), the sequences of feasible payoffs are of the form: θ times 0, then θ times 1, then 0 forever. One easily obtains for G_n:

$$\lim_{n \to \infty} v_n = \frac{1}{2}$$

(take θ to be near $n/2$). For G_λ, one maximizes $\sum_{m=\theta}^{2\theta-1} \lambda(1 - \lambda)^m = (1 - \lambda)^\theta (1 - (1 - \lambda)^\theta)$ w.r.t. θ, leading to $(1 - \lambda)^\theta \approx 1/2$ hence:

$$\lim_{\lambda \to 0} v_\lambda = \frac{1}{4}.$$

On the other hand on any feasible path the payoff is eventually 0 forever, hence one can consider the value V of the infinite game with payoff $\lim_{n\to\infty} \overline{g}_n$ and $V = 0$.

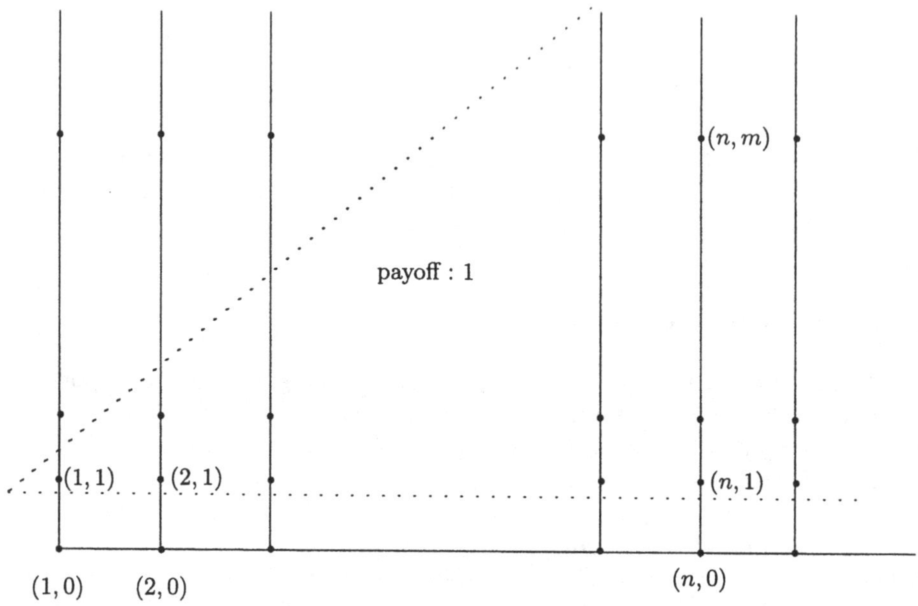

Fig. 1.2.

This example shows that, even when these various evaluations converge, they may differ. This question will be studied again in Chapter 5, Subsection 5.7.4.

1.4.2 Asymptotic approach and infinite game

The following MDP is also a recursive game (see Exercise 5.9.8): the payoff is 0 until it is absorbing. In particular $\lim_{n\to\infty} g_n = \lim_{n\to\infty} \overline{g}_n$ exists and this defines the payoff of the infinite game. We denote its value by V, if it exists. Explicitely, the set of states is $\Omega = \mathbb{N} \cup \{\partial\}$. From ∂ one chooses a point in $\mathbb{N}$ (the payoff is 0). Then the transition is deterministic from n to $n-1$ with payoff 0 until state 0 which is absorbing with payoff -1.
Clearly $v_n(\partial) = v_\lambda(\partial) = 0$ for all n and all λ (by choosing from ∂ a state in $\mathbb{N}$ large enough).
However on each feasible history the payoff is eventually -1. Hence V exists and is -1.

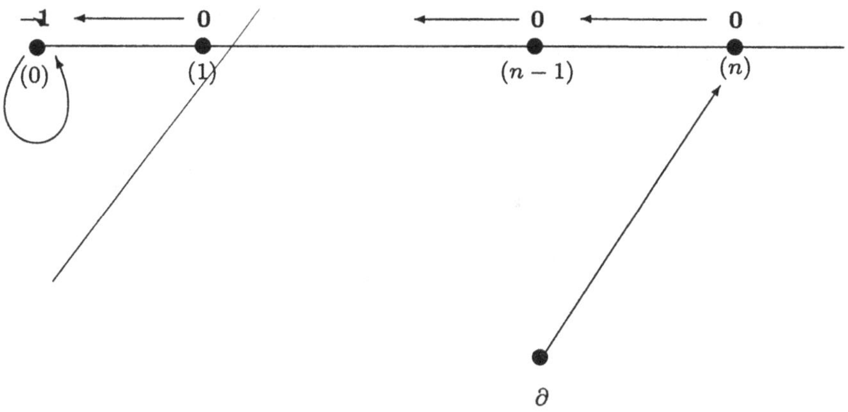

Fig. 1.3.

1.4.3 On stationary strategies

We consider again a MDP with state set Ω finite $= \{0, 1\}$ and moves set $\mathbb{N}$. The payoff in state 1 is 1. In state 0 it is 0 and absorbing.
Move n leads from state 1 to state 0 with probability $\frac{1}{n}$. Hence clearly any stationary strategy (which depends only on the current state hence here is constant) gives an asymptotic payoff 0 (the probability of reaching 0 is 1) while for any $\varepsilon > 0$ there exists a sequence n_ℓ such that the probability of absorption is less than $\sum_\ell \frac{1}{n_\ell} \leq \varepsilon$ and otherwise the payoff is 1.

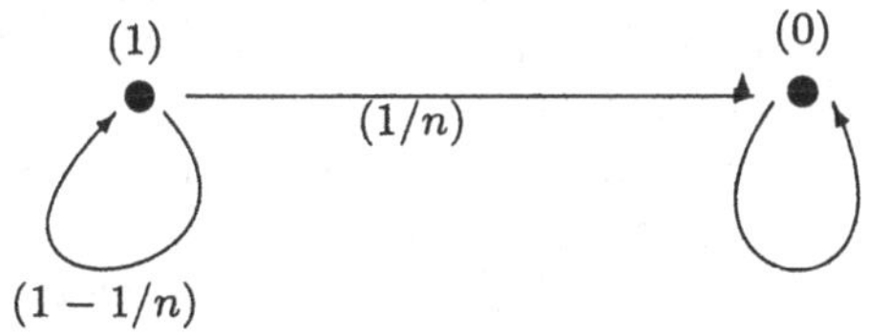

Fig. 1.4.

1.4.4 Compact move space and no optimal strategy

This example shows that when the strategy set is not finite optimal strategy may not exist.
Let $\Omega = \{0, -1, 1\}$ be the state space. In state ω the payoff is ω. States 1 and -1 are absorbing and the initial state is 0. For any choice $t \in [0, 1/2]$, the probability of moving to state 1 is t and to state -1 is t^2. The uniform value

of the infinite game is 1 but no strategy that reach state 1 can avoid reaching state -1 with positive probability.

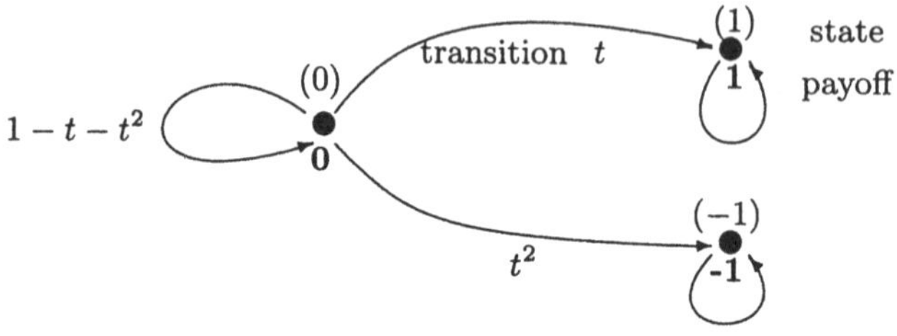

Fig. 1.5.

Note that this is again a recursive game.

1.4.5 The compact case

Two approaches for evaluating properties of a game repeated a large number of stages will be considered.
One is trough a family of games with (expected) duration increasing to ∞. For each game in the sequence the minmax theorem applies to obtain a value and the corresponding sequence of values will be studied.
The other point of view considers strategies in the infinite game and asks for uniform bounds on sequence of payoffs (Chapter 3, Section 3.1). We refer to it as the **uniform approach.**
In the first case (limit of finitely repeated games as the length goes to ∞ or limit of games with discount factor going to 0, for example) there is a natural embedding in a game of fixed duration while in the other case (infinite game) the past payoff is always negligible.
Formally a game in n stages where the payoff at stage m is g_m and the total payoff is $\overline{g}_n = \frac{1}{n}\sum_{m=1}^{n} g_m$ can be related to a game $\mathcal{G}$ being played in continuous time between time 0 and 1. The payoff at time t is g_t and the total payoff given by $\int_0^1 g_t dt$. However the strategies are requested to be constant on each interval of time $[\frac{m}{n}, \frac{m+1}{n})$.
Similarly, for a game with discount factor λ, hence with overall payoff $g_\lambda = \sum_{m=1}^{\infty} \lambda(1-\lambda)^{m-1} g_m$, the strategies are constant on each interval of time $[x_m, x_{m+1})$ with $x_0 = 0, x_1 = \lambda, ..., x_{m+1} = x_m + \lambda(1-\lambda)^m$. The weight of stage m in G_λ is its duration in $\mathcal{G}$. As n goes to ∞ or λ goes to 0, the size of each stage interval goes to 0.
Note that a similar embedding extends to any family of probability distributions $\{\mu_\alpha\}$ on the positive integers $\{1, ..., n, ...\}$. The payoff of the corresponding **generalized game** G_α is: $g_\alpha = \sum_m \mu_\alpha(m) g_m$. A natural requirement is that the weights are decreasing $\mu_\alpha(m+1) \leq \mu_\alpha(m)$ and then it is enough to

impose the condition $\mu_\alpha(1)\to 0$ for studying asymptotic properties.
In some cases we will exhibit a true "limit game" with payoff and strategies defined on the intervall $[0,1]$ and the previous finite or discounted versions will appear as discrete approximations (Chapter 5, Subsection 5.3.2).
This representation explain why this **asymptotic approach** will be also called the **compact case**.

1.5 Notes

The notion of game goes back to the 17th century but the introduction of mixed strategies is due to Borel (1921). von Neumann published his minmax theorem in 1928.
The examples in section 2 are due to Aumann and Maschler (1995), see also Zamir (1992). Section 3 follows Zamir (1973b). Example 4.1 is due to Lehrer and Sorin (1992). The literature on MDP is huge: see e.g. Dynkin and Yuskhkevich (1979) and Puterman (1994).
For an illuminating discussion of the different approaches to long term repeated games, see Aumann and Maschler (1995), Chapter 2, postscripts c, f and g.

A first approach to (strategic) game theory can be found in
Chapter 2 of : Basar T. and G.J. Olsder (1999) *Dynamic Noncooperative Game Theory*, SIAM Classics in Applied Mathematics.
Chapters 2 and 3 of : van Damme E. (1991) Stability and Perfection of Nash Equilibria, Springer .
Chapters 2 and 3 of : Myerson R. (1991) *Game Theory*, Harvard University Press.
Chapters 2 and 3 of : Osborne M. and A. Rubinstein (1994) *A Course in Game Theory*, M.I.T. Press.
Chapter II of : Owen G. (1995) *Game Theory*, Academic Press.
Chapters 1 and 2 of : Vorob'ev N.N. (1977) *Game Theory*, Springer.

2 Games with incomplete information

This chapter presents basic properties of a one stage game with incomplete information on one side . They will be applied in the sequel to the compact case (finitely or discounted repeated games), when using the normalized strategic form (Chapter 3, Section 3.1), and also to the infinite game (Chapters 3 and 6). Moreover some of these properties are relevant for the case of incomplete information on both cases (Chapter 4). This explains why rather than working with finite sets of strategies or simplexes over finite sets we consider general convex sets of strategies.

2.1 Presentation

Let K be a finite set. S and T are convex subsets of a topological vector space (locally convex Hausdorff). For each k, a two-person zero-sum game is defined by a payoff function G^k from $S \times T$ to $\mathbb{R}$: Player 1 (the **maximizer**) chooses s in S, Player 2 (the **minimizer**) chooses t in T, and the payoff is $G^k(s,t)$. G^k is bilinear and uniformly bounded: $\|G\|_\infty = \sup_{k,s,t} |G^k(s,t)| < \infty$.

To each p in the simplex over K, $\Delta(K)$, is associated a **game with lack of information on one side**, $G(p)$, played in two stages as follows:
- stage 0: k is chosen according to the probability p on K and communicated to Player 1 only (Player 1 is the **informed** player and Player 2 the **uninformed** one),
- stage 1: both players choose a move in their move sets, s in S and t in T, the payoff is $G^k(s,t)$.

Both players know the above description of the game.

The **strategic form** of the game $G(p)$ is represented by the triple $(G^p; S^K, T)$. Given the strategy $s = \{s^k\}_{k \in K}$ in S^K of Player 1 (where s^k is the move of Player 1 if k is announced) and the strategy t in T of Player 2 the payoff is:

$$G^p(s,t) = \sum\nolimits_{k \in K} p^k G^k(s^k, t).$$

Thus a strategy of Player 1 in the game G^p is a vector of moves.
The minmax and maxmin, as a function of p in $\Delta(K)$ are defined and denoted as follows.

Notation
$\overline{v}(p) = \inf_{t \in T} \sup_{s \in S^K} G^p(s,t)$
$\underline{v}(p) = \sup_{s \in S^K} \inf_{t \in T} G^p(s,t)$

A first easy but important property is the next one.

Proposition 2.1
$\underline{v}(p)$ and $\overline{v}(p)$ are Lipschitz with constant $\|G\|_\infty$ on $\Delta(K)$.

Proof
Note that the payoff functions in $G(p_1)$ and $G(p_2)$ differ by at most $\|G\|_\infty \|p_1 - p_2\|_1$. ∎

2.2 Concavity

A fundamental role is played by concavity in games with incomplete information. This reflects both the use of his information by the informed player (explicitly in Proposition 2.3) and the way the uninformed player deals with his uncertainty.

Proposition 2.2
$\overline{v}(p)$ and $\underline{v}(p)$ are concave on $\Delta(K)$.

Proof
The proof is based on the property that in a zero-sum framework, "the value of information is positive". In fact assume that Player 1 obtains more information: this increases his strategy set since information dependent strategies can be used. Hence both the maxmin and the minmax increase.

Let $p = \theta p_1 + (1-\theta)p_2$ with p, p_1, p_2 in $\Delta(K)$ and θ in $(0,1)$. Consider the two following games, G_1 and G_2, played in three stages. First i in $\{1,2\}$ is chosen according to the probability $(\theta, 1-\theta)$. Then the game is played as $G(p_i)$. In addition, in the game G_1, Player 2 knows i while he does not in G_2 (Player 1 is always told i). Remark that the game G_1 is equivalent to playing $G(p_1)$ with probability θ and $G(p_2)$ otherwise. On the other hand in the game G_2, the knowledge of i is irrelevant for Player 1 since he will know k, and from the point of view of Player 2 the overall probability of state k is p^k. The game G_2 is thus equivalent to $G(p)$. The game G_1 being more favourable to Player 2 (he has more information) the concavity inequality follows:

$$\theta v(p_1) + (1-\theta) v(p_2) \leq v(p)$$

where v stands for $\underline{v}$ or $\overline{v}$. ∎

Remarks
One could as well write explicitly, for example for the max min, first in the game G_1:

$$\underline{v}_1(p) = \max_{(s_1,s_2)\in S^K\times S^K} \min_{(t_1,t_2)\in T\times T}\{\theta\sum_k p_1^k G^k(s_1^k,t_1)+(1-\theta)\sum_k p_2^k G^k(s_2^k,t_2)\}$$

then in G_2:

$$\underline{v}_2(p) = \max_{(s_1,s_2)\in S^K\times S^K} \min_{t\in T}\{\theta\sum_k p_1^k G^k(s_1^k,t) + (1-\theta)\sum_k p_2^k G^k(s_2^k,t)\}.$$

For any $(s_1, s_2)\in S^K\times S^K$ there exists, by linearity of the payoff and convexity of S, $s\in S^K$ such that

$$p^k G^k(s^k,t) = \theta p_1^k G^k(s_1^k,t) + (1-\theta)G^k(s_2^k,t), \quad \forall k\in K, \quad \forall t\in T$$

so that

$$\underline{v}_2(p) = \underline{v}(p).$$

Note that the knowledge of the intermediate lottery p_i by Player 1 is crucial. The situation where Player 1 knows k and Player 2 knows p_i corresponds to a game with "lack of information on one and half sides" (see Chapter 4, Section 4.7.5) and has to be analyzed as a game with incomplete information on both sides.

An alternative proof of concavity of $\underline{v}(p)$ describes explicitly the use of his private information by the informed player.
This is the **splitting procedure**:

Proposition 2.3
Let L be a finite set and $p = \sum_{\ell\in L}\alpha_\ell p_\ell$, with α in $\Delta(L)$ and p, p_ℓ in $\Delta(K)$ for all ℓ in L.
Then there exists a transition probability μ from (K,p) to L such that:

$$P(\ell) = \alpha_\ell \quad \textit{and} \quad P(.|\ell) = p_\ell$$

where $P = p \bullet \mu$ is the probability induced by p and μ on $K\times L$: $P(k,\ell) = p^k\mu^k(\ell)$.

Proof
Let $\mu^k(\ell) = \alpha_\ell \dfrac{p_\ell^k}{p^k}$, for k in the support of p. Then $P(\ell) = \sum_{k\in K} p^k\mu^k(\ell) = \alpha_\ell$ and $P(k|\ell) = \dfrac{p^k\mu^k(\ell)}{\alpha_\ell} = p_\ell^k$. ∎

The interpretation is that μ (which will be a **signalling strategy** $\{\mu^k\}$ on the set L of signals) can generate any random variable with finitely many

values in $\Delta(K)$ and expectation p.
An easy consequence of the previous result is:

Corollary 2.4
$\underline{v}(p)$ is concave on $\Delta(K)$.

Proof
Let $p = \sum_{\ell \in L} \alpha_\ell p_\ell$, with α in $\Delta(L)$ and p, p_ℓ in $\Delta(K)$ for all ℓ. Let also s_ℓ in S^K be ε-optimal in $G(p_\ell)$. Introduce a strategy s for Player 1 in $G(p)$ as follows. Consider μ as defined in Proposition 2.3. If the state is k use the lottery μ^k, and if the outcome is ℓ play s_ℓ. explicitly s^k is the mixture $\sum_{\ell \in L} \mu^k(\ell) s^k_\ell$. To get a lower bound on Player 1's payoff assume even that Player 2 is informed upon ℓ: he is then facing strategy s_ℓ. By Proposition 2.3 this occurs with probability α_ℓ and the conditional probability on K is p_ℓ, hence the game is $G(p_\ell)$, so that:

$$\underline{v}(p) \geq \sum_{\ell \in L} \alpha_\ell \, \underline{v}(p_\ell) - \varepsilon, \quad \forall \varepsilon > 0.$$

■

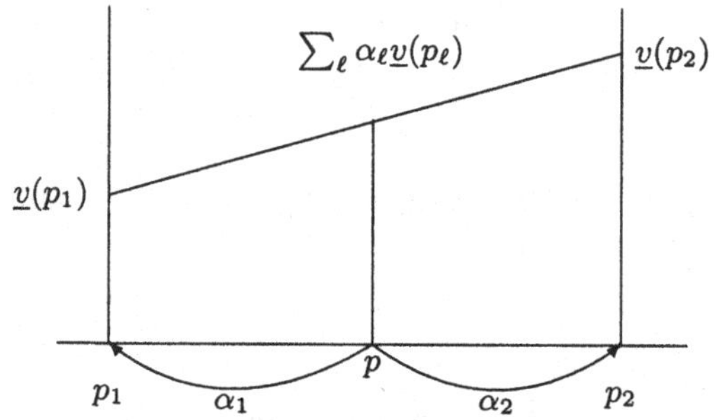

Fig. 2.1. The splitting procedure

The basic idea is twofold:
- the informed player follows a type dependent signalling strategy to realize the splitting procedure,
- by using as signals optimal strategies s_i in $G(p_i)$, Player 1 generates a random event such that Player 2 is facing s_i in $G(p_i)$, with probability α_i.

Another proof of concavity for $\overline{v}(p)$ follows from the following:

Proposition 2.5
For any t in T, $\sup_{S^K} G^p(s,t)$ is linear in p on $\Delta(K)$.

Proof
Since $s = \{s^k\}$, Player 1 can maximize on each factor, for each t, hence:

$$\sup_{s\in S^K} \sum p^k G^k(s^k,t) = \sum_k p^k \sup_{s^k\in S} G^k(s^k,t)$$

so that $\overline{v}$ is the infimum of linear functions on $\Delta(K)$, hence concave. ∎

2.3 Approachable vectors

To each strategy t of Player 2 one associates a vector $z(t)$ with: $z^k(t) = \sup_{s\in S} G^k(s,t)$. This corresponds to the **vector payoff** that Player 2 can guarantee by playing t.
Let $Z' = \{z(t); t\in T\} + \mathbb{R}_+^K$ be the upper comprehensive hull of $z(T)$ ($z\in Z'$ and $z' \geq z$ imply $z'\in Z'$) and $\overline{Z}'$ its closure.

Proposition 2.6
a) $\forall z\in Z', \forall q\in\Delta(K), \qquad \langle q,z\rangle \geq \overline{v}(q)$
b) Z' *is convex*
c) For each $q\in\Delta(K)$*, there exists* $z\in\overline{Z'}$ *such that:*

$$\langle q,z\rangle \ \leq \ \overline{v}(q).$$

Proof
a) Follows from Proposition 2.5.
b) Note that $z(\theta t_1 + (1-\theta)t_2) \ \leq \ \theta z(t_1) + (1-\theta)z(t_2)$.
c) If t is ε-optimal in $G(q)$, $\langle q, z(t)\rangle \ \leq \ \overline{v}(q)+\varepsilon$. ∎

Consider the set of "vectors above $\overline{v}$":

$$Z = \{z\in\mathbb{R}^K; \langle q,z\rangle \geq \overline{v}(q), \forall q\in\Delta(K)\}.$$

Proposition 2.7

$$Z = \overline{Z'}$$

$$\overline{v}(p) = \min_Z \langle p,z\rangle = \min_{Z_C} \langle p,z\rangle$$

where Z_C *is the intersection of* Z *with the cube* $[-\|G\|_\infty, +\|G\|_\infty]^K$.

Proof
By Proposition 2.6 *a)*, the inclusion $\overline{Z'}\subset Z$ holds.
Assume z in $Z\setminus\overline{Z'}$. Then z can be strongly separated from the closed convex set $\overline{Z'}$: there exists q in $\mathbb{R}^K$ and $\delta>0$ such that:

$$\langle q,z\rangle + \delta \ \leq \ \langle q,z'\rangle, \quad \forall z'\in Z'$$

Since Z' is upper comprehensive, q is positive and one can assume $q\in\Delta(K)$. Then one obtains:

$$\overline{v}(q) + \delta \leq \langle q, z \rangle + \delta \leq \langle q, z' \rangle, \quad \forall z' \in Z'$$

which contradicts Proposition 2.6 *c)*.
Finally, by definition $|z^k(t)| \leq \|G\|_\infty$, for all $k \in K$ and $t \in T$. ■

The duality between p and z is as follows:

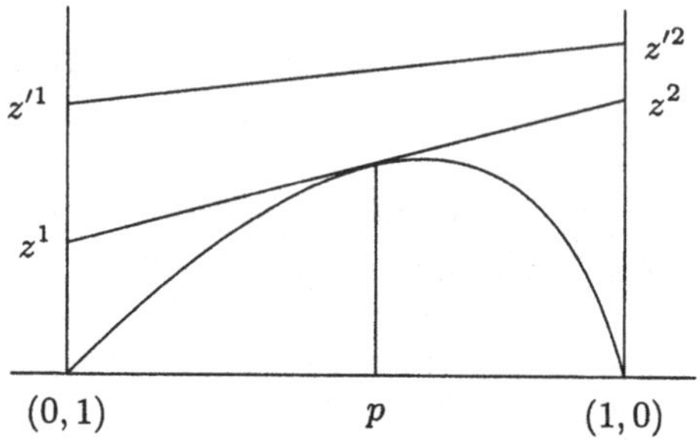

Fig. 2.2. The hyperplanes above $\overline{v}$

Z describes the convex set of hyperplanes above $\overline{v}$. If z is a supporting hyperplane to $\overline{v}$ at p, z belongs to the boundary of Z and p is a normal vector there:

$$\langle p, z \rangle \leq \langle p, z' \rangle \quad \forall z' \in Z.$$

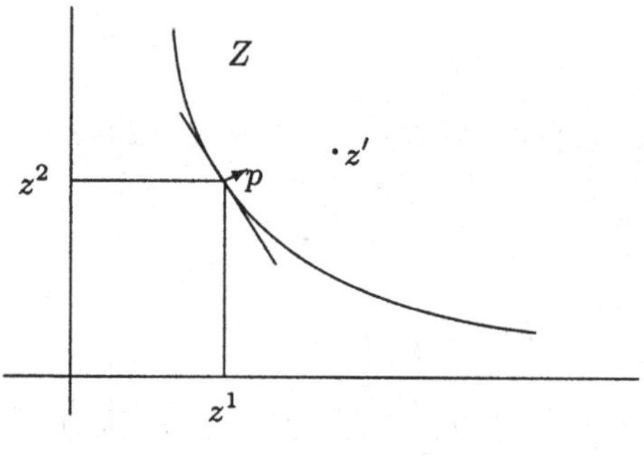

Fig. 2.3. The approachable vectors

Definition
Z is the set of **approachable vectors**. For any $z \in Z$ Player 2 can force a vector payoff in $z - \mathbb{R}_+^K$ since for any $\varepsilon > 0$ there exists $t \in T$ with $z(t) \leq z + \varepsilon$.

2.4 Dual game

For each vector x in $\mathbb{R}^K$, the **dual game** $G^*(x)$ is defined as follows: Player 1 chooses k; then both players play as in $G(p)$ at stage one, namely choose s in S and t in T; the payoff is $G^k(s,t) - x^k$.
The strategic form of $G^*(x)$ corresponds to the strategy set $K \times S$ for Player 1, T for Player 2 and the payoff function $h[x]$ on $(K \times S, T)$: $h[x](k,s;t) = G^k(s,t) - x^k$.
A mixed strategy π of Player 1 belongs to $\Delta(K \times S)$ and can be written as $\pi = p \bullet s$ in $\Delta(K) \times S^K$. explicitly $\pi(k,s) = p^k s^k$ where p is the marginal on K and s^k the conditional distribution on S, for each k (G^k being linear in s and S being convex, one identifies S and $\Delta(S)$).

Notation
$\overline{w}(x)$ and $\underline{w}(x)$ denote the minmax and maxmin of $G^*(x)$.

Note that $h[x] - h[y] \leq \|x - y\|_1$ and that $h[x+a] = h[x] - a$, where for a in $\mathbb{R}$, $x + a$ denotes the vector $\{x^k + a\}_{k \in K}$. Hence:

Proposition 2.8
$\overline{w}(x)$ and $\underline{w}(x)$ are Lipschitz with constant 1 *on* $\mathbb{R}^K$ *and satisfy* $f(x+a) = f(x) - a$.

The next result justifies the terminology "dual game" (we recall the notations of Appendix A.7).

Theorem 2.9

$$\underline{w}(x) = \max_{p \in \Delta(K)} \{\underline{v}(p) - \langle p, x \rangle\} = \Lambda_s(\underline{v})(x) \tag{2.1}$$

$$\underline{v}(p) = \inf_{x \in \mathbb{R}^K} \{\underline{w}(x) + \langle p, x \rangle\} = \Lambda_i(\underline{w})(p) \tag{2.2}$$

$$\overline{v}(p) = \inf_{x \in \mathbb{R}^K} \{\overline{w}(x) + \langle p, x \rangle\} = \Lambda_i(\overline{w})(p) \tag{2.3}$$

$$\overline{w}(x) = \max_{p \in \Delta(K)} \{\overline{v}(p) - \langle p, x \rangle\} = \Lambda_s(\overline{v})(x) \tag{2.4}$$

Proof
By definition:

$$\begin{aligned}\underline{w}(x) &= \sup_{p,s} \inf_t \{G^p(s,t) - \langle p, x \rangle\} \\ &= \sup_p \{\underline{v}(p) - \langle p, x \rangle\}\end{aligned}$$

and one can replace the sup by a max on $\Delta(K)$ since $\underline{v}$ is continuous. (2.2) is obtained by duality since $\underline{v}$ is concave and continuous on $\Delta(K)$, hence $\underline{v} = \Lambda_i \circ \Lambda_s(\underline{v})$ (Appendix A.7).

Let now t be an ε-optimal strategy of Player 2 in $G^*(x)$. Then:

$$G^k(s,t) - x^k \leq \overline{w}(x) + \varepsilon, \quad \forall s \in S, \quad \forall k \in K$$

so that $z(t) \leq x + \overline{w}(x) + \varepsilon$ and:

$$\overline{v}(p) \leq \inf_{x \in \mathbb{R}^K} \{\overline{w}(x) + \langle p, x \rangle\}$$

by Proposition 2.7. On the other hand if t is an ε-optimal strategy of Player 2 in $G(p)$, one has: $\overline{v}(p) \geq \langle p, z(t) \rangle - \varepsilon$. Finally, if Player 2 plays t in the game $G^*(z(t))$, all payoffs are less then 0, hence $\overline{w}(z(t)) \leq 0$. We obtain, for all p and $\varepsilon > 0$ the existence of some x with:

$$\overline{v}(p) \geq \overline{w}(x) + \langle p, x \rangle - \varepsilon$$

and (2.3) follows. Finally $\overline{w}(x)$ is convex (see Proposition 2.6 *b)*), hence (2.4) is obtained by duality. ∎

Remark

Equality in (2.1) implies that x and p are conjugate: $-p$ belongs to the subdifferential of the convex function $\overline{w}$ at x and x belongs to the superdifferential of the concave function $\overline{v}$ at p and similarly for equations (2.2), (2.3) and (2.4).

Note that the previous proof also showed:

Corollary 2.10

Given x, let p achieve the maximum in (2.1) and s be ε-optimal for Player 1 in $G(p)$. Then $p \bullet s$ is ε-optimal for Player 1 in $G^*(x)$.
Given p, let x achieve the infimum up to ε in (2.3). Then any t, ε-optimal for Player 2 in $G^*(x)$ is 2 ε-optimal for Player 2 in $G(p)$.

The next result relates approachability vectors and dual game. It translates the fact that from the point of view of Player 2 the underlying game is a game with "vector payoffs".

Corollary 2.11

$$\overline{w}(x) \leq 0 \iff x \in Z$$

$$\overline{v}(p) = \min_{x, \overline{w}(x) \leq 0} \langle p, x \rangle = \min_{x \in [-C, +C]^K} \{\overline{w}(x) + \langle p, x \rangle\}$$

Proof

Use Proposition 2.7 and 2.8 and the proof of Theorem 2.9. ∎

Comments
Note that since we are dealing with a one stage game, no aspect related to transmission of information (bluff or signalling) occurs. The properties obtained are only consequences of asymmetric uncertainty.

2.5 Notes

The main properties in Section 2 are due to Aumann and Maschler (1966); they were obtained in the framework of **repeated** games with incomplete information (see Chapters 3 and 4).
Approachable vectors are related to Blackwell's Theorem (Appendix B) and appear explicitly in MSZ (III.4) and in Mertens (1998).
The dual game was introduced by De Meyer (1998) in the framework of the study of the **error term** for repeated games with lack of information on one side (see Chapter 3, Section 3.7).

3 Repeated games with lack of information on one side

3.1 Presentation

The model of repeated game with incomplete information is introduced here in the simplest framework: lack of information on one side and standard signalling (some extensions will be discussed in Section 7 and others will be found in the next Chapter 4 and in Chapter 6).

As in the previous Chapter 2, a finite family of two person zero-sum games, G^k, k in K, is given. In addition each game G^k is finite and identified with a $I \times J$ matrix G^k. Let $\|G\| = \max_{i,j,k} |G^k_{ij}|$.
For each p in $\Delta(K)$ a **game form** $\mathcal{G}(p)$ is defined as follows:
- at stage 0, k is chosen according to the probability p on K and communicated to Player 1 only,
- at stage 1, Player 1 chooses a **move** i_1 in I, Player 2 chooses a move j_1 in J and the couple (i_1, j_1) is told to both,
- inductively, at stage m, knowing the past **history** $h_m = (i_1, j_1, \ldots, i_{m-1}, j_{m-1})$, Player 1 (resp. Player 2) chooses i_m in I (resp. j_m in J) and the new history $h_{m+1} = (h_m, i_m, j_m)$ is told to both.

Both players know the above description (public knowledge).

$H_m = (I \times J)^{m-1}$ is the set of histories at stage m (H_1 is reduced to one point $= \{\emptyset\}$) and $H = \cup_{m \geq 1} H_m$ is the set of all histories.
$S = \Delta(I)$ and $T = \Delta(J)$ denote the sets of **mixed moves** of the players.
A **behavioral strategy** (see Appendix D) or simply a strategy for Player 1 is a map σ from $K \times H$ to S. Explicitly $\sigma^k(h)[i]$ is the probability that Player 1 will play the move i, given his **type** (corresponding to his private information) k and the history h. The notation σ_m will also be used for the restriction of σ to $K \times H_m$: it describes the behavior of Player 1 at stage m. Similarly, but taking into account his lack of information on k, a strategy for Player 2 is a map τ from H to T. τ will also be written as the sequence $\{\tau_m\}$, where τ_m is the restriction of τ to H_m.
Σ and $\mathcal{T}$ denote the sets of strategies of Player 1 and Player 2, respectively. Pure strategies are defined similarly by maps with values in the sets I and J of moves.

A triple (p, σ, τ) in $\Delta(K) \times \Sigma \times \mathcal{T}$ induces a probability distribution $P_{p,\sigma,\tau}$ on the set $H_\infty = K \times (I \times J)^\infty$ of **plays** (endowed with the σ-field $\mathcal{H}_\infty = \mathcal{K} \vee_{m \geq 1} \mathcal{H}_m$ where $\mathcal{K}$ is the discrete field on K and $\mathcal{H}_m$ is generated by the

cylinders above H_m, see Appendix D). $E_{p,\sigma,\tau}$ stands for the corresponding expectation. As usual, k is identified with the Dirac mass on k so that $E_{p,\sigma,\tau}$ is also the average $\sum_{k\in K} p^k E_{k,\sigma^k,\tau}$.
Each play $\xi = (k, i_1, j_1, \ldots, i_m, j_m, \ldots)$ defines a sequence of payoffs $\{g_m\}$ where $g_m = G^k_{i_m j_m}$ is the random payoff at stage m and $\gamma^p_m(\sigma,\tau) = E_{p,\sigma,\tau}(g_m)$ is its expectation.

Several games are associated to the game form $\mathcal{G}(p)$. They differ in the way the stream of payoffs is evaluated.

a) $G_n(p)$ is the n-**stage game** with payoff function:

$$\overline{\gamma}^p_n(\sigma,\tau) = \frac{1}{n}\sum\nolimits_{m=1}^{n} \gamma^p_m(\sigma,\tau) = E_{p,\sigma,\tau}(\overline{g}_n)$$

where $\overline{g}_n = \frac{1}{n}\sum_{m=1}^{n} g_m$. Note that $G_n(p)$ is actually a finite game (finitely many pure strategies) since the behavior after stage n is irrelevant for the payoff. Let $v_n(p)$ denote its value. We will mainly study asymptotic properties of the sequence $\{v_n\}$ as n goes to ∞.

b) $G_\lambda(p)$ is the λ-**discounted game** with payoff function:

$$\gamma^p_\lambda(\sigma,\tau) = \sum\nolimits_{m=1}^{\infty} \lambda(1-\lambda)^{m-1}\gamma^p_m(\sigma,\tau) = E_{p,\sigma,\tau}(g_\lambda)$$

where $g_\lambda = \sum_{m=1}^{\infty}\lambda(1-\lambda)^{m-1} g_m$. Consider here Σ and $\mathcal{T}$ as sets of **mixed strategies** (see Appendix D), namely probabilities on the sets of pure strategies (with the cylindar σ-field). The stage payoff being uniformly bounded, $\gamma^p_\lambda(\sigma,\tau)$ is jointly continuous for the product topology for which Σ and $\mathcal{T}$ are compact. In addition γ^p_λ is bilinear hence (Theorem A. 11) $G_\lambda(p)$ has a value $v_\lambda(p)$. Its behavior as λ goes to 0 will be examined.

Remark
All the results of Chapter 2 apply to $v_n(p)$ and $v_\lambda(p)$ by considering the **strategic form** of the game: the strategy sets are Σ and $\mathcal{T}$ and the payoff is $\overline{\gamma}^p_n$ or γ^p_λ.

c) Finally $G_\infty(p)$ is the **infinitely repeated game** where the payoff is not explicitly determined on plays but minmax and maxmin are defined through uniform properties as follows.

Definitions
Player 1 can **guarantee** an amount $f(p)$ in $G_\infty(p)$ if:
$\forall\varepsilon > 0, \exists\sigma$, strategy of Player 1 and N such that $\forall\tau$, strategy of Player 2 and $\forall n \geq N$:

$$\overline{\gamma}^p_n(\sigma,\tau) \;\geq\; f(p) - \varepsilon.$$

This corresponds to a strong property: up to any ε Player 1 can obtain f uniformly w.r.t. time and strategy of the opponent.

Player 2 can **defend** an amount $f(p)$ in $G_\infty(p)$ if:
$\forall\varepsilon > 0, \forall\sigma\in\Sigma, \exists\tau\in\mathcal{T}$ and $\exists N$ such that, $\forall n \geq N$:

$$\overline{\gamma}_n^p(\sigma,\tau) \leq f(p) + \varepsilon.$$

This statement is stronger than the negation of the previous one since the "best reply" is uniform in time.
Dual definitions are obtained by exchanging the players, explicitely:
Player 2 can **guarantee** an amount $f(p)$ in $G_\infty(p)$ if:
$\forall\varepsilon > 0, \exists\tau$ and N such that $\forall\sigma$ and $\forall n \geq N$:

$$\overline{\gamma}_n^p(\sigma,\tau) \leq f(p) + \varepsilon.$$

Player 1 can **defend** an amount $f(p)$ in $G_\infty(p)$ if:
$\forall\varepsilon > 0, \forall\tau, \exists\sigma$ and $\exists N$ such that, $\forall n \geq N$:

$$\overline{\gamma}_n^p(\sigma,\tau) \geq f(p) - \varepsilon.$$

$\underline{V}(p)$ is the **maxmin** of $G_\infty(p)$ if Player 1 can guarantee $\underline{V}(p)$ and Player 2 can defend it.
A similar requirement holds for the **minmax** $\overline{V}(p)$: Player 2 can guarantee it and Player 1 can defend it.
$G_\infty(p)$ has a value if $\underline{V}(p) = \overline{V}(p)$, it is denoted $v_\infty(p)$ and called the **uniform value**.

Recall that γ_λ^p is a convex combination of $\{\overline{\gamma}_n^p\}$. More generally for any probability distribution μ on the positive integers with $\mu(m)$ decreasing one has:

$$g_\mu = \sum_{m=1}^{\infty} \mu(m) g_m = \sum_{m=1}^{\infty}\sum_{\ell=m}^{\infty} [\mu(\ell) - \mu(\ell+1)] g_m$$

$$= \sum_{\ell=1}^{\infty} [\mu(\ell) - \mu(\ell+1)][\sum_{m=1}^{\ell} g_\ell] = \sum_{m=1}^{\infty} \ell\, [(\mu(\ell) - \mu(\ell+1)]\, \overline{g}_\ell \qquad (*)$$

hence g_μ is a convex combination of $\overline{g}_\ell$.
Thus, an immediate consequence of the above definitions is:

Lemma 3.1
If Player 1 can guarantee $f(p)$, then $\liminf_{n\to\infty} v_n(p)$ and $\liminf_{\lambda\to 0} v_\lambda(p)$ are greater than $f(p)$. Dual properties hold for Player 2.
If $v_\infty(p)$ exists, then:

$$v_\infty(p) = \lim_{n\to\infty} v_n(p) = \lim_{\lambda\to 0} v_\lambda(p).$$

Similar notations and definitions will be used for other classes of repeated games: games with lack of information on both sides, stochastic games,... (see Chapters 4, 5 and 6). In all cases Lemma 3.1 holds.

The remaining of this chapter is as follows:
Section 2 presents basic properties related to the transmission of information by Player 1: martingale of posterior probabilities and non-revealing strategies.
Section 3 deals with asymptotic results based on the fact that Player 2 can "restart" the game at any stage.
Section 4 is central and proves the main results concerning the compact case: existence and characterization of the limit of the values for the finite and discounted games G_n and G_λ.
Several alternative approaches to the results of Section 4 are studied in Section 5.
Section 6 concerns the infinite game G_∞ and proves the existence of v_∞.
Some extensions and results, without proofs, are presented in Section 7.

3.2 Basic properties

3.2.1 The martingale of posteriors

Consider a vector s in $S^K = \Delta(I)^K$. This corresponds to a one stage strategy of the informed Player 1. Together with p in $\Delta(K)$, s induces a transition probability from K to I: let k be chosen according to $p\in\Delta(K)$, then i be selected following s^k. This generates a posterior probability on K taking, with probability $\bar{s}(i) = \sum_k p^k s_i^k$, the value $p(i)\in\Delta(K)$, where $p^k(i) = Prob(k|i) = \dfrac{p^k s_i^k}{\bar{s}(i)}$.
Note that $p = p(i)$ for all i, if and only if $s^k = \bar{s}$ for all k with $p^k > 0$. Hence the vector s is **non-revealing** at p if it belongs to:

$$NR(p) = \{s\in S^K;\ \forall k, k' \quad p^k p^{k'} > 0 \Rightarrow s^k = s^{k'}\}.$$

Observe that $NR(p)$ is a convex polyhedron.

Definition
$D(p)$ is the **non-revealing game** played on $NR(p)\times T$ with payoff $\sum_k p^k s^k G^k t$.
By Theorem A.11 it has a value, denoted by $u(p)$ and note that:

$$u(p) = \mathtt{val}_{S\times T}\ s(\sum\nolimits_k p^k G^k)t.$$

In particular u is $\|G\|$ Lipschitz.

Introduce now a measure of the **revelation** contained in s at p by:

$$R(s,p) = E(\|s^k - \overline{s}\|_1) = \sum_{k\in K} p^k \|s^k - \overline{s}\|_1$$

and the classical one stage **variation of the martingale** of posteriors :

$$V(s,p) = E(\|p(i) - p\|_1) = \sum_{i\in I} \overline{s}_i \|p(i) - p\|_1.$$

Hence s is non-revealing at p if and only if $R(s,p)$ is 0 or equivalently if $V(s,p)$ is 0. In fact a much more precise relation holds, both quantities being a measure of correlation of the distribution induced by p and s on $K\times I$: $\pi = p \bullet s$ with $\pi(k,i) = p^k s^k(i)$.

Proposition 3.2

$$R(s,p) = V(s,p).$$

Proof

$$R(s,p) = \sum_{k\in K} p^k \|s^k - \overline{s}\|_1 = \sum_{k\in K} p^k \sum_{i\in I} |s_i^k - \overline{s}_i|$$

$$= \sum_{i\in I} \overline{s}_i \sum_{k\in K} |p^k(i) - p^k| = \sum_{i\in I} \overline{s}_i \|p(i) - p\|_1 = V(s,p).$$

∎

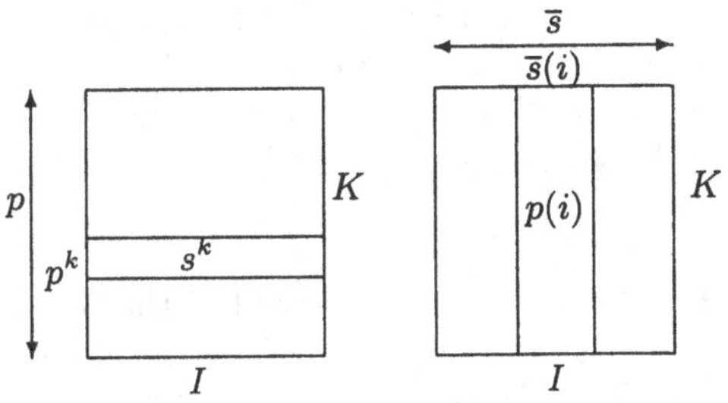

Fig. 3.1. Two measures of correlation of $\pi = p \bullet s$ on $K\times I$

We apply now the previous results to the repeated game.
Given a pair of strategies (σ, τ), consider the distribution induced by $P_{p,\sigma,\tau}$ on $K\times H_m$. The conditional distribution on K given $h_m \in H_m$ is the **posterior distribution** p_m at stage m, with $p_1 = p$. It corresponds to the beliefs of the uninformed Player 2 on the type of Player 1.
To describe explicitely the law of p_m, define the **average strategy** of Player 1 at h_m by:

$$\overline{\sigma}(h_m) = \sum_{k\in K} p_m^k(h_m)\sigma^k(h_m).$$

Then one has:

Proposition 3.3
For any (σ, τ), the sequence $\{p_m\}$ is a $\mathcal{H}_m$-martingale on H with values in $\Delta(K)$.
Explicitely, for h_m in H_m:

$$p^k_{m+1}(h_m, i_m, j_m) = p^k_m(h_m)\frac{\sigma^k(h_m)[i_m]}{\overline{\sigma}(h_m)[i_m]}.$$

The sequence of moves can be represented as a tree and p_m is defined on the nodes.

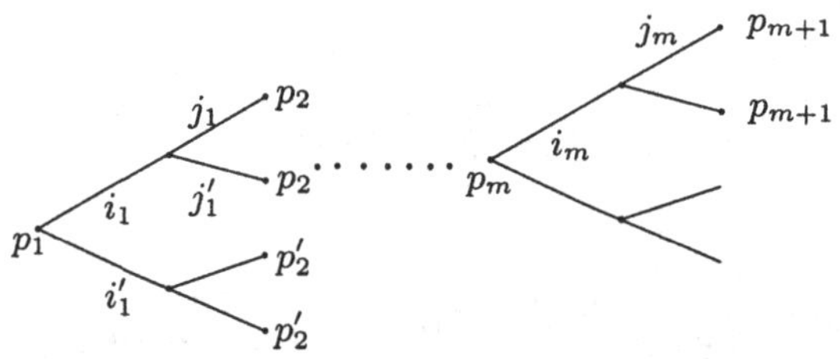

Fig. 3.2. The history tree and the martingale of posteriors

Note that p_{m+1} depends only on p_m and i_m, but that the distribution of $\{p_m\}$ depends upon the moves of Player 2 through Player 1's strategy.
Proposition 3.3 corresponds to the "direct part" of the splitting procedure in Proposition 2.3: any strategy generates a martingale (and any martingale - with finitely many values - can be generated).

Definitions
Given σ and τ, the **local game** at h_m is defined by the posterior $p_m(h_m) \in \Delta(K)$, the type dependent mixed move of Player 1, $\sigma(h_m) \in \Delta(I)^K$, and the mixed move of Player 2, $\tau(h_m) \in \Delta(J)$. It describes the "state" of the game and the current behavior of the players after history h_m.
The **local payoff**

$$\rho_m(\sigma, \tau)(h_m) = \sum\nolimits_k p^k_m(h_m)\sigma^k(h_m)G^k\tau(h_m)$$

is the payoff in the local game at h_m, i.e. the conditional expected payoff at stage m, given h_m.
The **local revelation** of σ, given p at h_m is defined by

$$LR(\sigma, p)(h_m) = R(\sigma(h_m), p_m(h_m)).$$

It is an evaluation of the use of information by Player 1 playing σ, at stage m after history h_m.
The **local variation** of p given σ at h_m is:

$$LV(\sigma,p)(h_m) = V(\sigma(h_m), p_m(h_m)).$$

This quantity is thus the conditional expected one stage variation in L_1-norm of the martingale of posteriors at stage m after history h_m.
A strategy σ is **non-revealing at** h_m if $\sigma(h_m)$ belongs to $NR(p_m(h_m))$ and then $p_{m+1}(h_m,\cdot) = p_m(h_m)$. σ is **non-revealing** if it is non-revealing after each history. When following a non-revealing strategy Player 1 is not using his information (on k) or equivalently not transmitting any information: the posterior does not change.

3.2.2 Variation of bounded martingales

We prove here some results on bounded martingales that will be used in the next section.
Let $(\Omega,\mathcal{F},Q)$ a probability space, $\mathcal{F}_m$ be a filtration on $\mathcal{F}$ and $\mathbf{q} = \{q_m\}$ be a $\mathcal{F}_m$-martingale with values in $[0,1]$ starting from $q_1 = E(q_m)$. Its total variation in L_2 norm:

$$\mathbf{V}^2(\mathbf{q}) = E(\sum\nolimits_{m=1}^{\infty}(q_{m+1}-q_m)^2)$$

is uniformly bounded:

Lemma 3.4

$$\mathbf{V}^2(\mathbf{q}) \le q_1(1-q_1).$$

Proof
$\{q_m\}$ being a martingale, its increments are uncorrelated:

$$E\left((q_{m+1}-q_m)(q_{\ell+1}-q_\ell)\right) = E(E((q_{m+1}-q_m)(q_{\ell+1}-q_\ell)|\mathcal{F}_m))$$
$$= E\left(E((q_{m+1}-q_m)|\mathcal{F}_m)(q_{\ell+1}-q_\ell)\right) = 0, \text{ for } m > \ell.$$

Hence:

$$E\left(\sum\nolimits_{m=1}^{M}(q_{m+1}-q_m)^2\right) = E\left([\sum\nolimits_{m=1}^{M}(q_{m+1}-q_m)]^2\right)$$
$$= E\left((q_{M+1}-q_1)^2\right) \le q_1(1-q_1).$$

The majorant being uniform in M, the result follows. ∎

The above bound allows to control the following quantities expressed in L_1 norm.
Define:

$$\mathbf{V}^1_n(\mathbf{q}) = E\left(\sum\nolimits_{m=1}^{n}|q_{m+1}-q_m|\right)$$

and

$$\mathbf{v}^1_\lambda(\mathbf{q}) = E\left(\sum\nolimits_{m=1}^{\infty} \lambda(1-\lambda)^{m-1}|q_{m+1} - q_m|\right).$$

Lemma 3.5

$$\mathbf{V}^1_n(\mathbf{q}) \leq \sqrt{q_1(1-q_1)}\sqrt{n}$$

$$\mathbf{v}^1_\lambda(\mathbf{q}) \leq \sqrt{q_1(1-q_1)}\sqrt{(\frac{\lambda}{2-\lambda})}.$$

Proof
By Cauchy-Schwartz inequality:

$$\begin{aligned}\mathbf{V}^1_n(\mathbf{q}) &\leq E\left([\sum\nolimits_{m=1}^{n} (q_{m+1} - q_m)^2]^{1/2}\sqrt{n}\right) \\ &\leq [E\left(\sum\nolimits_{m=1}^{n} (q_{m+1} - q_m)^2\right)]^{1/2}\sqrt{n}\end{aligned}$$

and similarly:

$$\mathbf{v}^1_\lambda(\mathbf{q}) \leq [E\left(\sum\nolimits_{m=1}^{\infty} (q_{m+1} - q_m)^2\right)]^{1/2}[\sum\nolimits_{m=1}^{\infty} \lambda^2(1-\lambda)^{2(m-1)}]^{1/2}$$

then use Lemma 3.4. ∎

3.3 Preliminary results

The **repeated non-revealing game** $\mathcal{D}(p)$ is the game $\mathcal{G}(p)$ where Player 1 is restricted to non-revealing strategies. Since the posterior distribution is constant this game is equivalent to the repetition of the one shot game $D(p)$ on $NR(p) \times T$ with stage payoff $\sum_{k \in K} p^k s^k G^k t$ and value $u(p)$. This quantity is a lower bound of what Player 1 can achieve, namely:

Proposition 3.6
Player 1 has a strategy σ such that for any strategy τ of Player 2 and $n \geq 1$:

$$\gamma^p_n(\sigma, \tau) \geq u(p).$$

Proof
Let Player 1 play s i.i.d. where s is an optimal strategy in $D(p)$. ∎

The next results follow from the fact that Player 2 can always forget the past history, hence "restart" the game from the beginning, since he did not transmit any information. Explicitly, given any past history, Player 2 can obtain $v_m(p)$ as average payoff for the m next stages and $v_\lambda(p)$ as future λ-discounted payoff.

Proposition 3.7
Player 2 can guarantee $\liminf_{n\to\infty} v_n(p)$.

Proof
Let n_ℓ be a sequence going to ∞ on which $v_{n_\ell}(p)$ converge to $\liminf_{n\to\infty} v_n(p)$ and let τ_ℓ be optimal for Player 2 in $G_{n_\ell}(p)$.
Let τ be the following strategy: use τ_1 for the n_1 first stages, then use τ_1 again but applied to histories from stage n_1+1 on (i.e. Player 2 forgets the past history up to stage n_1) for the following n_1 stages and so on n_2 times. Inductively use τ_ℓ successively on $n_{\ell+1}$ blocks of size n_ℓ. On each block of length n_ℓ, the average expected payoff of Player 2 is at most $v_{n_\ell}(p)$: in fact for Player 1 to play past blocks dependent strategy is useless. Formally, if τ is independent of h then σ is payoff equivalent to its expectation w.r.t. h. The result follows since $\overline{\gamma}_n$ is an average of v_{n_ℓ}, for ℓ between 1 and some $L(n)$, up to an error term bounded by $\dfrac{\|G\|}{L(n)}$.

∎

Corollary 3.8
$\lim_{n\to\infty} v_n$ *exists.*

Proof
Use Lemma 3.1 (for Player 2) and Proposition 3.7. ∎

Proposition 3.9

$$\forall\lambda\in(0,1)\quad \limsup_{n\to\infty} v_n \le v_\lambda.$$

Proof
Let τ_λ be an optimal strategy of Player 2 in $G_\lambda(p)$. Let τ be the following strategy: use τ_λ at stage 1 and inductively at each stage use the previously followed strategy with probability $(1-\lambda)$ and restart τ_λ with probability λ. Explicitely, at stage m given the history h_m, $\tau(h_m)$ is the mixture: $\tau_\lambda(h_m)$ with probability $(1-\lambda)^{m-1}$ and $\tau_\lambda(h_m^r)$ with probability $\lambda(1-\lambda)^{r-1}$, $r=1,\ldots,m-1$, where h_m^r corresponds to the $(r-1)$ last moves $(i_{m-r+1}, j_{m-r+1}, \ldots, i_{m-1}, j_{m-1})$ of the history h_m.
The average payoff $\overline{\gamma}_n$ is thus a convex combination of discounted payoffs evaluated between stage m and n and where Player 2 uses τ_λ from stage m on. For a given λ any such payoff, for which $n-m$ is large enough, is atmost $v_\lambda(p)$. Formally, let N such that $(1-\lambda)^N \le \varepsilon$. Then for $n \ge N/\varepsilon$ one has: $v_n(p) \le v_\lambda(p) + 2\varepsilon\|G\|$.

∎

Corollary 3.10
$\lim_{\lambda\to 0} v_\lambda$ *exists and equals* $\lim_{n\to\infty} v_n$.

Proof
Follows from Lemma 3.1, Proposition 3.7 and Proposition 3.9. ∎

An independent proof of existence of $\lim_{\lambda \to 0} v_\lambda$ follows from the next property:

Proposition 3.11
v_λ is decreasing in λ.

Proof
For any sequence $\{a_n\}$ let:

$$F_\lambda^m(\{a_n\}) = \sum_{n=1}^{\infty} \lambda(1-\lambda)^{n-1} a_{m+n}$$

be its discounted evaluation from stage $m+1$ on. Assume $\mu < \lambda$, then one has

$$F_\mu^0(\{a_n\}) = \frac{\mu}{\lambda} F_\lambda^0(\{a_n\}) + \frac{\mu}{\lambda}(\lambda - \mu) \sum_{m=0}^{\infty} (1-\mu)^m F_\lambda^{m+1}(\{a_n\})$$

Let τ_λ be optimal in $G_\lambda(p)$ and note that, for any σ and any $m \geq 0$:

$$F_\lambda^m(\{\gamma_n^p(\sigma, \tau_\lambda)\}) = \sum_{n=1}^{\infty} \lambda(1-\lambda)^{n-1} \gamma_{n+m}^p(\sigma, \tau_\lambda) \leq v_\lambda(p)$$

because by restarting τ_λ from stage $m+1$ on Player 2 could get $v_\lambda(p)$.
Let then Player 2 use τ_λ in the game $G_\mu(p)$. Since $\gamma_\mu^p(\sigma, \tau_\lambda)$ belongs to the convex hull of the family $F_\lambda^m(\{\gamma_n^p(\sigma, \tau_\lambda)\})$ the result follows from the previous upper bound.

∎

3.4 Asymptotic approach

Recall that all results from Chapter 2 concerning the maxmin $\underline{v}(p)$ or the minmax $\overline{v}(p)$ of a game with incomplete information apply here to $v_n(p)$ and $v_\lambda(p)$. In particular these functions are concave and $\|G\|$ Lipschitz.
Also, as far as the value is concerned, the minmax theorem allows to assume both strategies to be known by both players, hence p_m is also known by both players and will play the role of a **state variable**.

Notation
Given a real function f on $\Delta(K)$, its **concavification**, denoted $\mathtt{Cav}\ f$, is the smallest function, concave and greater than f on $\Delta(K)$ (see App. A.8).

3.4.1 The finite case

The first basic result is a lower bound on the values:

Proposition 3.12

$$v_n \geq \text{Cav } u, \qquad \forall n \geq 1.$$

Proof
From Proposition 3.6, there exists σ such that $\overline{\gamma}_n^p(\sigma, \tau) \geq u(p)$ for all τ. Hence v_n is greater than u. Use Proposition 2.2 to conclude. ∎

In words, the informed player can obtain u by not using (or revealing) his information and then can concavify by using it (splitting procedure).

To get an upper bound on the value, fix σ. Then define a "behavioral reply" of Player 2 in the local game at h_m. Let us compare the corresponding local payoff $\rho_m(\sigma, \tau)(h_m)$ to the one obtained with the local average (nonrevealing) strategy $\overline{\sigma}(h_m)$:

Lemma 3.13
For any σ, τ and any h_m:

$$|\sum_k p_m^k(h_m)\sigma^k(h_m)G^k\tau(h_m) - \sum_k p_m^k(h_m)\overline{\sigma}(h_m)G^k\tau(h_m)|$$
$$\leq \|G\| LR(\sigma, p)(h_m).$$

Proof
Follows from the definitions of $\overline{\sigma}(h_m)$ and $LR(\sigma, p)$ (Section 2.1). ∎

This lemma is crucial: it shows that the gain that Player 1 can obtain by using his information is bounded by (a constant time) the local revelation of its strategy. Recall that this quantity equals the local variation of the martingale of posteriors and that the total n-stage variation is of the order of $\sqrt{n}$. The upper bound on the payoffs will then follow.

Define a strategy $\tau' = \tau'(\sigma)$ of Player 2 as: play at h_m an optimal strategy in $D(p(h_m))$, namely optimally in the one stage game where the parameter is $p(h_m)$ and none of the players is informed. Then one has:

Proposition 3.14
For any σ, if player 2 uses $\tau' = \tau'(\sigma)$

$$\rho_m(\sigma, \tau')(h_m) \leq u(p_m(h_m)) + \|G\| LV(\sigma, p)(h_m).$$

Proof
Note that $\overline{\sigma}(h_m)$ belongs to $NR(p_m(h_m))$, hence by the choice of τ':

$$\sum_k p_m^k(h_m)\overline{\sigma}(h_m)G^k\tau(h_m) \leq u(p_m(h_m)).$$

The result then follows from Proposition 3.2 and the previous Lemma 3.13.

■

We now obtain an upper bound for v_n, using the results from section 2.2.

Proposition 3.15

$$v_n(p) \leq \mathtt{Cav}\ u(p) + \|G\|\frac{\sum_{k\in K}\sqrt{p^k(1-p^k)}}{\sqrt{n}}.$$

Proof
By taking expectation in Proposition 3.14 and then summation, one gets:

$$n\overline{\gamma}_n(\sigma,\tau') \leq \sum_{m=1}^{n}\left[E(u(p_m(h_m))) + \|G\|E(LV(\sigma,p)(h_m))\right].$$

Jensen's inequality gives:

$$\sum_{m=1}^{n} E(u(p_m(h_m))) \leq \sum_{m=1}^{n} E(\mathtt{Cav}\ u(p_m(h_m)))$$
$$\leq \sum_{m=1}^{n} \mathtt{Cav}\ u(E(p_m(h_m))) \leq n\ \mathtt{Cav}\ u(p).$$

For the other term one has, by applying Lemma 3.5 to each component $\{p_m^k\}$:

$$\sum_{m=1}^{n} E(LV(\sigma,p)(h_m)) = \sum_{m=1}^{n} E(\|p_{m+1}-p_m\|_1)$$
$$= \sum_k \mathbf{V}_n^1(\mathbf{p}^k) \leq \sum_k \sqrt{p^k(1-p^k)}\sqrt{n}.$$

Hence this proves that, for any strategy σ of Player 1, there exists a strategy τ' of Player 2 such that:

$$n\overline{\gamma}_n(\sigma,\tau') \leq n\ \mathtt{Cav}\ u(p) + \|G\|\sum_k \sqrt{p^k(1-p^k)}\sqrt{n}.$$

■

The main result of this section is summarized in the following:

Theorem 3.16
There exist a constant C $(= \#K\|G\|)$ such that:

$$\mathtt{Cav}\ u(p) \leq v_n(p) \leq \mathtt{Cav}\ u(p) + \frac{C}{\sqrt{n}}.$$

The value of the n stage game converges from above to the concavification of the value of the non-revealing game, at a speed bounded by $0(\frac{1}{\sqrt{n}})$.

3.4.2 The discounted case

The game with discounted payoffs $G_\lambda(p)$ is now considered. The proofs are exactly similar to the previous ones. One first obtains:

Proposition 3.17

$$v_\lambda(p) \geq \mathtt{Cav}\ u(p).$$

Proof
As in Proposition 3.12, the proof follows from Proposition 3.6 and Proposition 2.2. ∎

The use of the same strategy $\tau' = \tau(\sigma)$ as in the previous section 4.1 leads to:

Proposition 3.18

$$v_\lambda(p) \leq \mathtt{Cav}\ u(p) + \|G\| \sum\nolimits_{k\in K} \sqrt{p^k(1-p^k)} \frac{\sqrt{\lambda}}{\sqrt{2-\lambda}}.$$

Proof
By taking expectation in Proposition 3.14 and then summation, one obtains:

$$\gamma_\lambda(\sigma,\tau') \leq \sum\nolimits_{m=1}^{\infty} \lambda(1-\lambda)^{m-1}[E(u(p_m(h_m))) + \|G\| E(LV(p)(h_m))].$$

Jensen's inequality gives:

$$\begin{aligned}\sum\nolimits_{m=1}^{\infty} \lambda(1-\lambda)^{m-1} E(u(p_m(h_m))) &\leq \sum\nolimits_{m=1}^{\infty} \lambda(1-\lambda)^{m-1} E(\mathtt{Cav}\ u(p_m(h_m)))\\ &\leq \sum\nolimits_{m=1}^{\infty} \lambda(1-\lambda)^{m-1} \mathtt{Cav}\ u(E(p_m(h_m)))\\ &\leq \mathtt{Cav}\ u(p).\end{aligned}$$

For the remaining term one writes:

$$\begin{aligned}\sum\nolimits_{m=1}^{\infty} \lambda(1-\lambda)^{m-1} E(LV(p)(h_m)) &= \sum\nolimits_{m=1}^{\infty} \lambda(1-\lambda)^{m-1} E(\|p_{m+1}-p_m\|_1)\\ &= \sum\nolimits_{k} \mathbf{v}_\lambda^1(\mathbf{p}^k) \leq \sum\nolimits_{k} \sqrt{p^k(1-p^k)} \frac{\sqrt{\lambda}}{\sqrt{2-\lambda}},\end{aligned}$$

by applying Lemma 3.5 to each component $\{p_m^k\}$ of the martingale $\{p_m\}$. ∎

The counterpart of Theorem 3.16 is the following

Theorem 3.19
There exists a constant $C(= \#K\|G\|)$ such that:

$$\text{Cav } u(p) \le v_\lambda(p) \le \text{Cav } u(p) + C\frac{\sqrt{\lambda}}{\sqrt{2-\lambda}}.$$

Comments

The same result extends to other versions of the compact case, (Recall Chapter 1, Section 1.3.5). Let μ^α be a family of probability distributions on $\mathbb{N}^*$ with $\mu^\alpha(n) \ge \mu^\alpha(n+1)$ and $\mu^\alpha(1) \to 0$ as $\alpha \to 0$. Then $\sum_n \mu^\alpha(n)\gamma_n$ is a convex combination of the averages $\overline{\gamma}_n$. Hence with the same proof (including the same local strategy of Player 2) one obtains, with obvious notations, that v_α converges to Cav u at a speed bounded by $0(\mu^\alpha(1)^{1/2})$.

3.5 Recursive structure and operator approach

In this section the structural connection between the initial game and the game after one stage is explicitely used. (This is a specific case of general recursive properties, see MSZ (Chapter IV.3)). A similar analysis is done in Chapter 4, Section 4.6, Chapter 5 and Appendix C, Section 4.

3.5.1 Recursive formula

The $n+1$ stage game can be decomposed into a one stage game and a game of length n starting with a new value of the state variable. Similarly the discounted game gives raise to a one shot game followed by a (normalized) discounted game with a new state variable. The relations between the values are as follows:

Proposition 3.20

v_n and v_λ satisfy:

$$(n+1)v_{n+1}(p) = \text{val}_{S^K \times T}\{\sum_{k\in K} s^k p^k G^k t + n\sum_{i\in I} \overline{s}(i) v_n(p(i))\}$$

$$v_\lambda(p) = \text{val}_{S^K \times T}\{\lambda\sum_{k\in K} s^k p^k G^k t + (1-\lambda)\sum_{i\in I} \overline{s}(i) v_\lambda(p(i))\}$$

where $\overline{s} = \sum_{k\in K} p^k s^k$ and $p(i)$ is the conditional distribution on K given i (Section 2.1).

Proof

The proof is the same for both expressions.

Firstly, the concavity of v_n implies that $\sum_{i\in I}\overline{s}(i)v_n(p(i))$ is concave as a function of s on S^K. In fact if $s = \theta s_1 + (1-\theta)s_2$ with s, s_1 and s_2 in S^K and θ in $(0,1)$, one has:

$$\theta\overline{s}_1(i)v_n(p_1(i)) + (1-\theta)\overline{s}_2(i)v_n(p_2(i)) \leq \overline{s}(i)v_n(p(i))$$

with $p_\ell^k(i) = \dfrac{p^k s_\ell^k(i)}{\overline{s}_\ell(i)}$, $\ell = 1, 2$.
Hence by Theorem A.7, the game on $S^K \times T$ with payoff $\sum_k s^k p^k G^k t + n\sum_i \overline{s}(i)\ v_n(p(i))$ has a value. Let us prove that its maxmin equals $(n+1)\ v_{n+1}(p)$.
It is clear that by playing first some s that achieves the maximum, then optimally in $G_n(p(i))$, Player 1 obtains as payoff in $G_{n+1}(p)$ at least the expression on the right hand side. Explicitely, replacing the strategy of Player 2 from stage 2 on by its average conditionally on the first move of Player 1 does not change the payoff.
On the other hand, given any σ, let $s = \sigma_1$ be its component at stage 1 and choose t realizing the minimum given s. The strategy τ consisting in playing t then optimally in $G_n(p(i))$, given the move i of Player 1, induces a total payoff in $G_{n+1}(p)$ less than the right hand side expression.

∎

We will sometimes write the above recursive formula in a compact way as :

$$(n+1)v_{n+1}(p) = \mathtt{val}\{g_1(p) + nE[v_n(\widetilde{p})]\}$$
$$v_\lambda(p) = \mathtt{val}\{\lambda g_1(p) + (1-\lambda)E[v_\lambda(\widetilde{p})]\}$$

The previous proof also proved properties of optimal strategies of Player 1, namely:

Corollary 3.21
Player 1 has an optimal strategy in $G_n(p)$ that depends only, at each stage m, on m and p_m. (This strategy is Markov on $\Delta(K)$).
Player 1 has an optimal strategy in $G_\lambda(p)$ that depends only at each stage m on p_m. (This strategy is stationary on $\Delta(K)$).
In particular both are independent of the moves of Player 2.

The recursive formula (3.20) for v_n allows also to get the next property:

Proposition 3.22
The sequence v_n is decreasing in n.

Proof
The proof, by induction, relies on the concavity of v_n and on the above recursive formula. Start with:

$$2v_2(p) = \mathtt{val}\{g_1(p) + E[v_1(\widetilde{p})]\}$$
$$\leq \mathtt{val}\{g_1(p) + v_1(p)\} \leq 2v_1(p)$$

by Jensen's inequality.
Similarly assuming $v_n \leq v_{n-1}$ one obtains:

$$\begin{aligned}(n+1)\, v_{n+1}(p) &= \mathtt{val}\{g_1(p) + nE[v_n(\widetilde{p})]\} \\ &\leq \mathtt{val}\{g_1(p) + (n-1)E[v_{n-1}(\widetilde{p})] + v_n(p)\} \\ &\leq \mathtt{val}\{g_1(p) + (n-1)E[v_{n-1}(\widetilde{p})]\} + v_n(p) \\ &= (n+1)\, v_n(p).\end{aligned}$$

■

We now use the dual approach (Chapter 2.4) to obtain results first dealing with the dual values, then concerning optimal strategies in the dual game. Recall how the dual game $G_n^*(x)$ of $G_n(p)$ (resp. $G_\lambda^*(x)$ of $G_\lambda(p)$) is played. Player 1 chooses k (Player 2 is not informed). Then at each stage m, knowing the previous history h_m, both players choose moves (i_m, j_m). The payoff is $\frac{1}{n}\sum_{m=1}^n G^k_{i_m j_m} - x^k$ (resp. $\sum_{m=1}^\infty \lambda(1-\lambda)^{m-1} G^k_{i_m j_m} - x^k$). The value $w_n(x)$ of $G_n^*(x)$ satisfies:

$$w_n(x) = \max_{p \in \Delta(K)} \{v_n(p) - \langle p, x\rangle\} = \Lambda_s(v_n)(x)$$

and similarly for $G_\lambda^*(x)$:

$$w_\lambda(x) = \max_{p \in \Delta(K)} \{v_\lambda(p) - \langle p, x\rangle\} = \Lambda_s(v_\lambda)(x)$$

From the "primal" recursive formula one obtains:

Proposition 3.23

$$w_{n+1}(x) = \min_{t \in T} \max_{i \in I} (\frac{n}{n+1}) w_n(\frac{n+1}{n} x - \frac{1}{n} G_i t)$$

$$w_\lambda(x) = \min_{t \in T} \max_{i \in I} (1-\lambda) w_\lambda(\frac{1}{1-\lambda} x - \frac{\lambda}{1-\lambda} G_i t)$$

where $G_i t$ is the vector $\{G_i^k t\}$ in $\mathbb{R}^K$ with component $G_i^k t = \sum_{j \in J} G^k_{ij} t_j$ and $w_0(x) = \max_{k \in K}\{-x^k\}$.

Proof
Using the recursive formula for v_{n+1} one has:

$$w_{n+1}(x) = \max_{p \in \Delta(K)} \max_{s \in S^K} \min_{t \in T}$$

$$\left\{\frac{1}{n+1}[\sum_k s^k p^k G^k t + n \sum_i \overline{s}(i) v_n(p(i))] - \langle p, x\rangle\right\}.$$

Let $\pi = p \bullet \sigma$ in $\Delta(K \times I)$ (see Section 2.1) with $\pi(k,i) = p^k s^k(i)$. (Note that the range is all $\Delta(K \times I)$).
One has, π^i being the marginal on I, π_k the marginal on K and $\pi[i]$ the conditional on K given i:

$$w_{n+1}(x) = \max_{\pi \in \Delta(K \times I)} \min_{t \in T}$$

$$\left\{ \frac{1}{n+1} [\sum\nolimits_{k,i} \pi(k,i) G_i^k t + n \sum\nolimits_i \pi^i v_n(\pi[i])] - \sum\nolimits_k \pi_k x^k \right\}.$$

As in Proposition 3.20 one shows here that $\sum_i \pi^i v_n(\pi[i])$ is concave in π. Hence one obtains:

$$w_{n+1}(x) = \min_{t \in T} \max_{\pi \in \Delta(K \times I)}$$

$$\left\{ \frac{1}{n+1} [\sum\nolimits_{k,i} \pi(k,i) G_i^k t + n \sum\nolimits_i \pi^i v_n(\pi[i])] - \sum\nolimits_k \pi_k x^k \right\}.$$

Using $\pi_k = \sum_i \pi^i \pi[i]^k$, this gives:

$$w_{n+1}(x) = \min_{t \in T} \max_{\pi^i \in \Delta(I)} \max_{\pi[i] \in \Delta(K)}$$

$$\left\{ \frac{n}{n+1} \sum\nolimits_i \pi^i \left(v_n(\pi[i]) - \langle \pi[i], \frac{n+1}{n} x - \frac{1}{n} G_i t \rangle \right) \right\}.$$

Thus, by taking extreme points in the max on $\Delta(I)$:

$$w_{n+1}(x) = \min_{t \in T} \max_{i \in I} (\frac{n}{n+1}) \max_{p \in \Delta(K)} \left(v_n(p) - \langle p, \frac{n+1}{n} x - \frac{1}{n} A_i t \rangle \right)$$

and the result follows.
The proof is similar for w_λ.

∎

In this framework, the role of **state variable** is played for Player 2 by the sequence $\{x_m\}$ in $\mathbb{R}^K$, with $x_1 = x$ and inductively, in G_n^*:

$$n x_1 = (n-1) x_2 + G_{i_1} t_1$$

$$(n-1) x_2 = (n-2) x_3 + G_{i_2} t_2, \ldots$$

where i_m is the move of Player 1 and t_m is the mixed move of Player 2 at stage m
and in G_λ^*:

$$x_1 = (1-\lambda) x_2 + \lambda G_{i_1} t_1$$

$$x_2 = (1-\lambda) x_3 + \lambda G_{i_2} t_2, \ldots$$

Corollary 3.24
Player 2 has an optimal strategy in $G_n^*(x)$ that depends only, at each stage

m, on m and x_m (Markov on $\mathbb{R}^K$).
Player 2 has an optimal strategy in $G^*_\lambda(x)$ that depends only at each stage m on x_m (stationary on $\mathbb{R}^K$).
In particular both are independent of his own previous moves.

Proof
Consider $G^*_{n+1}(x)$ and t optimal for $w_{n+1}(x)$, following Proposition 3.23. Define the strategy τ of Player 2 as: using t at stage one and then given the move i of Player 1, playing optimally in $G^*_n(\frac{n+1}{n}x - \frac{1}{n}G_i t)$. We obtain, for any σ:

$$E_{\sigma\tau}(g_2 + \dots + g_{n+1} - n(\frac{n+1}{n}x - \frac{1}{n}G_i t)|i) \le n w_n(\frac{n+1}{n}x - \frac{1}{n}G_i t)$$

hence:

$$E_{\sigma\tau}(g_1 + g_2 \dots + g_{n+1} - (n+1)x) \le n \max_i w_n(\frac{n+1}{n}x - \frac{1}{n}G_i t).$$

Thus by the choice of t:

$$\begin{aligned}\frac{1}{n+1}E_{\sigma\tau}(g_1 + \dots + g_{n+1} - (n+1)x) &\le \frac{n}{n+1}\min_t \max_i w_n(\frac{n+1}{n}x - \frac{1}{n}G_i t)\\ &\le w_{n+1}(x).\end{aligned}$$

∎

We use now the relation between optimal strategies for Player 2 in the primal and dual games to obtain:

Corollary 3.25
Player 2 has an optimal strategy in $G_n(p)$ that depends only, at each stage m, on m and $(i_1, \dots, i_{m-1})$.
Player 2 has an optimal strategy in $G_\lambda(p)$ that depends only at each stage m on $(i_1, \dots, i_{m-1})$.
In particular both are independent of his own moves.

Proof
Follows from Corollary 2.10 and Corollary 3.24. Explicitely, given p choose x realizing $v_n(p) = \Lambda_i(w_n)(p)$, and play optimal in $G^*_n(x)$ as above.

∎

3.5.2 Cav u through the dual values

This section uses the recursive formula to obtain an alternative proof of Theorems 3.16 and 3.19 on the convergence of the values to Cav u in the

compact case.
Cav u appears here as the biconjugate of the function u (see Appendix C.8). The advantage of using the recursive formula for the dual is that, for example in the case of v_n, the factor $1/n$ is now inside the function w_n and this allows for a local expansion around x, recall Proposition 3.23.
Extend u by $-\infty$ outside $\Delta(K)$ and define:

$$\Lambda_s(u)(x) = \max_{p \in \mathbb{R}^K} (u(p) - \langle p, x\rangle) = \max_{p \in \Delta(K)} (u(p) - \langle p, x\rangle) = u^{\#}(x)$$

and

$$\Lambda_i \circ \Lambda_s(u)(p) = \min_{x \in \mathbb{R}^K} (\Lambda_s(u)(x) + \langle p, x\rangle).$$

Recall that, with these notations, $w_n = \Lambda_s(v_n)$ and $v_n = \Lambda_i(w_n)$ and similarly for w_λ and v_λ. Also for any u.s.c. function f on $\Delta(K)$ (extended outside by $-\infty$), one has $\Lambda_i \circ \Lambda_s(f) = \mathtt{Cav}\ f$ (Appendix A.8).
Note that both Λ_s and Λ_i are non expansive operators hence one has:

Proposition 3.26

$$\|w_n - \Lambda_s(u)\| = \|v_n - \mathtt{Cav}\ u\|$$
$$\|w_\lambda - \Lambda_s(u)\| = \|v_\lambda - \mathtt{Cav}\ u\|.$$

Thus the convergence of v_n or v_λ to Cav u will be a consequence of the convergence of the dual values to $u^{\#}$. An heuristic argument is as follows: Consider the operator defining the dual recursive formula in Proposition 3.23, written as $w_{n+1} = \Psi_{n+1}(w_n)$. One obtains for a C^2 function f:

$$\Psi_{n+1}(f)(x) - f(x) = \min_{t \in T} \max_{s \in S} \left[\frac{n}{n+1} f(x + \frac{1}{n}(x - sGt))\right] - f(x)$$

$$= \min_{t \in T} \max_{s \in S} \frac{1}{n+1} [-f(x) + \langle \nabla(f)(x), x - sGt\rangle] + 0(\frac{1}{n^2})$$

where sGt stands for the vector $\{sG^k t\}$. This leads to the differential equation:

$$-f(x) + \langle \nabla(f)(x), x\rangle + \mathtt{val}_{S\times T}\langle -\nabla(f)(x), sGt\rangle = 0. \qquad (E)$$

Assume $u^{\#}$ being C^2. Let q be the gradient of $u^{\#}$ at some point x. A consequence of Fenchel's duality gives (Proposition A.18):

$$u^{\#}(x) = u(-q) + \langle q, x\rangle$$

hence $-q = p$ is in the range $\Delta(K)$ of u. Since $u(p) = \mathtt{val}_{S\times T}\langle p, sGt\rangle$, it follows that $f = u^{\#}$ satisfies (E) so that $\Psi_{n+1}(u^{\#}) - u^{\#} = 0(\frac{1}{n^2})$.
The properties on approximate fixed points (Appendix C. 3), with $a = 1$ give the result.

For the proof, let $f_n(x)$ be the expectation $E(u^{\#}(x+\frac{z}{\sqrt{n}}))$ where the random variable z follows a centered and reduced normal law $\mathcal{N}(0,1)$. Then one has:

Proposition 3.27
There exist constants C_1 and C_2 such that:

$$\|w_n(x) - u^{\#}(x)\| \leq \frac{C_1}{\sqrt{n}}$$

$$\|w_\lambda(x) - u^{\#}(x)\| \leq C_2\sqrt{\lambda}.$$

Proof
The proof follows from Appendix C.3 by showing that f_n satisfy:

$$\|\Psi_{n+1}(f_n) - f_{n+1}\| \;\leq\; L_1 n^{-3/2}$$

and

$$\|u^{\#} - f_n\| \;\leq\; L_2 n^{-1/2}$$

for some L_1 and L_2 and similar properties for Ψ_λ, see Exercise 8.1. ∎

3.5.3 Cav u through non expansive mappings

The purpose of this section is to give a third alternative proof of Theorems 3.16 and 3.17 on the convergence of the values to Cav u in the compact case by using general properties of non expansive mappings and concavity.

From Proposition 3.20 one has $nv_n = \boldsymbol{\Psi}^n(0)$ where $\boldsymbol{\Psi}$ is a non expansive mapping given by:

$$\boldsymbol{\Psi}(f)(p) = \texttt{val}\ \{g_1(p) + E[f(\tilde{p})]\}.$$

Using Theorem C.1 one first deduces that $\|v_n\|_\infty$ converges.
Substracting for each k the constant α^k from the matrix G^k changes the value $v_n(p)$ to $v_n(p) - \langle p, \alpha\rangle$. Adding a constant to obtain non negative payoffs, one deduces that the sequence $w_n(\alpha) = \Lambda_s(v_n)(\alpha)$ converges to some $w(\alpha)$. Note that all functions $w_n(\alpha)$ are 1−Lipschitz hence the convergence is uniform on the cube $[-\|G\|, +\|G\|]^K$, hence v_n converges uniformly to $\Lambda_i(w) = v$ by Theorem 2.9.
It remains to identify the limit v. Let $(p_0, v(p_0))$ an extreme point of (the hypograph of) v. Let us prove that $u(p_0) = v(p_0)$: since v is concave, continuous and above u this will imply that $v =$ Cav u. Adding a linear form one can assume that $\|v\|_\infty = v(p_0) > v(p)$, for $p \neq p_0$. By Theorem C.1 again there exists ℓ_x in B', unit ball of the dual of the space of continuous functions on $\Delta(K)$, with:

$$\langle \ell_x, \Psi^n(x) - x \rangle \geq n\rho$$

and $\rho = \|v\|_\infty$. In particular taking for x the function 0 we obtain the existence of μ_0, a Borel measure on $\Delta(K)$ satisfying, for all n:

$$\int_{\Delta(K)} v_n(p)\mu_0(dp) \geq \rho$$

hence at the limit:

$$\int_{\Delta(K)} v(p)\mu_0(dp) \geq \rho$$

which implies that μ_0 is the Dirac mass at p_0, say δ_0.
Note that, for an alternative x, since $\|\Psi^n(x)/n - v_n\|$ goes to 0, one has also from Theorem C.1:

$$\langle \ell_x, v \rangle \geq \rho$$

hence $\ell_x = \delta_0$ as well.
Choose for x a concave, piecewise linear function, maximal at p_0 with $|x(p) - x(p_0)| \geq (2\|G\|/\varepsilon)\|p - p_0\|_1$. It follows that, in the evaluation of $\Psi(x)$, if the one stage variation of p is larger than ε, the loss in x cannot be compensated by a gain in g. Using Proposition 3.2 and Lemma 3.13 one thus obtains:

$$\Psi(x)(p_0) \leq u(p_0) + x(p_0) + 2\|G\|\varepsilon$$

so that

$$u(p_0) + 2\|G\|\varepsilon \geq \Psi(x)(p_0) - x(p_0) \geq \langle \ell_x, \Psi(x) - x \rangle \geq \rho \; = \; v(p_0)$$

hence the result.
For the discounted case, Theorem C.1.a gives the convergence of $\|\lambda x_\lambda\|_\infty = \|v_\lambda\|_\infty$. Now the concavity of v_λ implies, like for v_n, that the family v_λ converges to some function w. In addition, from Theorem C.1.b one has:

$$\langle \ell_0, w \rangle \leq \rho$$

so that $w(p_0) = \|w\|_\infty = u(p_0)$, hence w coincide with v at each extreme point and is below any supporting hyperplane to v, hence the equality.

∎

Comment
A fourth alternative proof of Theorems 3.16 and 3.19 uses the operator approach (Appendix C.4) and extends to the case of lack of information on both sides, see Chapter 4, Subsection 4.6.2.

3.6 Infinite game

The purpose of this section is to study the infinite game G_∞: value and optimal strategies.

3.6.1 Existence of v_∞

Using the definitions of Section 1 we first show:

Proposition 3.28
Player 1 can guarantee Cav u.

The proof follows from the two next properties:

Lemma 3.29
Player 1 can guarantee u.

Proof
Use Proposition 3.6. (Remark that the result is much stronger since it says that Player 1 can obtain at least u at each stage.) ∎

Lemma 3.30
If Player 1 can guarantee $f(p)$ *in* $G_\infty(p)$, *for all* p *in* $\Delta(K)$, *he can guarantee* Cav$f$.

Proof
The proof is similar to the proof of Corollary 2.4, using the splitting procedure, Proposition 2.3. ∎

Note that the previous results allow to define an optimal strategy of Player 1 having the stronger property of being independent of ε.
On the other hand one has:

Proposition 3.31
Player 2 can guarantee Cav u.

Proof
Use Proposition 3.7 and Theorem 3.16. ∎

This statement does not describe explicitely how Player 2 can guarantee this amount: the properties of v_n were obtained through a kind of "best reply" of Player 2 and no optimal strategy was exhibited.
Concluding from Proposition 3.28 and Proposition 3.31, we obtain the following existence property.

Theorem 3.32
v_∞ *exists and equals* Cav u.

3.6.2 An optimal strategy for Player 2

In fact a much more precise result is available, by constructing inductively an optimal strategy of Player 2 in $G_\infty(p)$. It relies on approachability properties, see Appendix B.
Consider the game with vector payoffs defined by the collection of matrices $\{G^k\}, k \in K$. Rather than evaluating his payoff according to some probability in $\Delta(K)$, Player 2 will take into consideration a vector payoff, corresponding to all possible values of k and approach a set in $\mathbb{R}^K$.
Introduce Z_∞, the set of vectors above u:

$$Z_\infty = \{z \in \mathbb{R}^K; \langle q, z \rangle \geq u(q), \forall q \in \Delta(K)\}$$

and for z in $\mathbb{R}^K$, let $M(z) = z - \mathbb{R}^K_+$ be the negative orthant translated by z.

Theorem 3.33

$$z \in Z_\infty \iff \text{Player 2 can approach } M(z).$$

Proof
$M(z)$ is convex hence we use Corollary B.4: $M(z)$ is approachable by Player 2 iff, for all α in $\mathbb{R}^K$:

$$\max_{t \in T} \min_{s \in S} \langle \alpha, sGt \rangle \geq \min_{c \in M(z)} \langle \alpha, c \rangle.$$

Since $M(z)$ is lower comprehensive, we can assume that all components of α are non positive and $\alpha \neq 0$, so that by normalizing one obtains $q = -\alpha$ in $\Delta(K)$ and:

$$\max_t \min_s \langle -q, sGt \rangle \geq -\langle q, z \rangle$$

which is:

$$\langle q, z \rangle \geq \min_t \max_s \langle q, sGt \rangle = u(q)$$

and this achieves the proof. ■

Corollary 3.34
There exists τ and a constant L such that for all σ:

$$\overline{\gamma}^p_n(\sigma, \tau) \leq \texttt{Cav}\ u(p) + \frac{L}{\sqrt{n}}.$$

Proof
Let z in Z_∞ such that:

$$\langle z, p \rangle = \texttt{Cav}\ u(p).$$

The set $M(z)$ is a **B**-set for Player 2. Thus, there exists a strategy τ and a constant L such that:

$$E_{\sigma,\tau}(d(\overline{g}_n, M(z))) \le \frac{L}{\sqrt{n}}.$$

So that, a fortiori, for each k:

$$\overline{\gamma}_n^k(\sigma,\tau) \le z^k + \frac{L}{\sqrt{n}},$$

hence

$$\overline{\gamma}_n^p(\sigma,\tau) = \sum\nolimits_k p^k \overline{\gamma}_n^k(\sigma,\tau) \le \langle z, p\rangle + \frac{L}{\sqrt{n}}$$

which ends the proof. ∎

3.7 Comments and extensions

3.7.1 Comments

a) Some properties obtained in this chapter have to be underlined.

First the analysis of the game cannot be done for a specific value of p without taking into consideration the whole range of p, $\Delta(K)$. This is the natural state space where the game, viewed as a stochastic game, see Chapter 5, evolves.

Second, it is amazing to get the value of a quite complex optimization problem (recall the uniformity property in the definition of the minmax and maxmin) explicitely as a function of the basic data; namely the value u of the one shot non-revealing game.

Note the deep disymmetry between the players: Player 1 can make a computation based on the maximal information he is transmitting. In the actual play of the game he cannot monitor nor observe the information or deduction of Player 2 but due to the zero-sum aspect, his computation gives a lower bound on his payoff.
On the other hand, even if Player 1's optimal strategy is unique, a computation of posteriors and a best reply behavior by Player 2, based on the assumption that Player 1 is playing optimally would be subject to bluffing. This is the reason why an optimal play of Player 2 cannot rely on conditional probabilities but on actual vector payoffs.

In G_∞, the optimal strategies have very specific features:
The informed player computes a splitting that allows to reach $\mathtt{Cav}\ u$ at p. He performs a type dependent lottery that achieves it and plays an optimal strategy in the non-revealing game at the realized posterior. Hence Player 1 is using once for all his information at the beginning of the game. It is as if

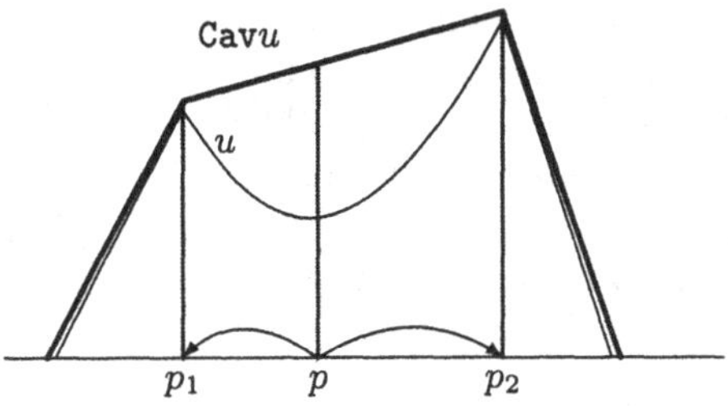

Fig. 3.3. The splitting strategy of Player 1

he was concentrating at the first stage all the informational content of his strategy.
On the other hand Player 2 computes a vector payoff defining a supporting hyperplane to Cav u at p and uses then Blackwell's approachability strategy for the corresponding set. Hence he also uses p once for all at stage one.

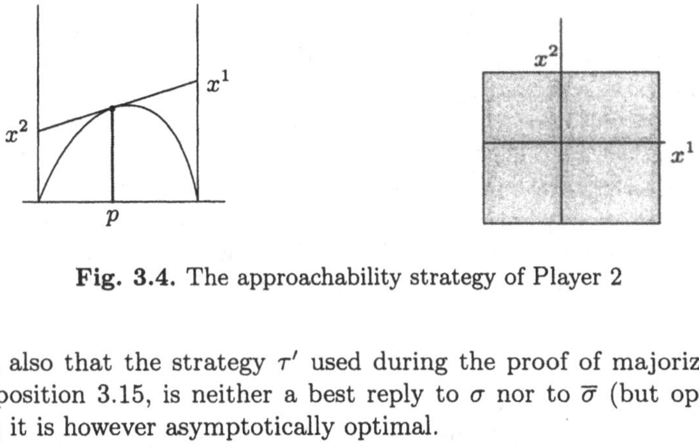

Fig. 3.4. The approachability strategy of Player 2

Remark also that the strategy τ' used during the proof of majorization of v_n, Proposition 3.15, is neither a best reply to σ nor to $\overline{\sigma}$ (but optimal in $D(p_m)$); it is however asymptotically optimal.

b) Back to the examples of Chapter 1, Section 1.1, one has the following features:
case 1 : $u(p) = p(1-p)$ is already concave hence equal to v_∞. The asymptotic value of information is 0 and the only optimal strategy is non revealing.
case 2 : $u(p) = -p(1-p)$ is convex hence Cav $u = 0$. Moreover the only optimal splitting is to the extreme points 0 and 1: completely revealing.
case 3 : $u(0) = 0$, $u(1/4) = 1$ and $u(1/2) = 0$. u is linear on each of these subintervals and $u(1/2+x) = u(1/2-x)$. Hence Cav u is 1 on $[1/4, 3/4]$ and the only optimal splitting at 1/2 is towards $\{1/4, 3/4\}$: it is partially revealing.

c) The relation between v_n and v_λ is quite complex.
v_n is mathematically simpler (piecewise linear, see Chapter 4 Ex.4.8.8) while

v_λ may not be derivable on a countable set (Exercise 8.2).
v_n is defined inductively while v_λ relies on a fixed point argument (but can be approximated by the value of the finitely repeated discounted game).
Concerning monotonicity note that the property for v_λ is general, Proposition 3.11, while the one for v_n uses the recursive formula, Proposition 3.22. As a matter of fact, in games with signals, see Subsection 7.4, v_n is not necessarily monotonic, Lehrer (1987), Yariv (1997).

d) The case where each player's initial information can be described by a partition $\mathcal{K}_i$ of K reduces to the present framework whenever $\mathcal{K}_1$ refines $\mathcal{K}_2$. In fact it is enough to redefine K as the set of atoms of $\mathcal{K}_1$ and to take as payoff the corresponding average.

3.7.2 The game on $[0, 1]$

(Chapter 1, Section 1.4.5)
a) The analysis in Section 4 leads to the following heuristics concerning the compact game in continuous time.
The total gain obtained by Player 1 by using his information is, for example in the n stage game, bounded by a quantity of the order of $\sqrt{n}$, hence the impact per stage vanishes as n goes to ∞. This means that essentially, as far as the payoff is concerned, Player 1 is playing in the non revealing game $D(p_m)$, at stage m. The "revealing" aspect of his strategy (somehow the variance) determines the martingale p_m, the mean is thus a free variable that can be chosen to play optimally, hence we can assume that Player 1 obtains $u(p_m)$ at each stage. Since Proposition 3.14 shows exactly the counterpart for Player 2, one can look at the compact game as a **splitting game** with payoff:

$$\int_0^1 u(p_t)dt$$

where p_t is a martingale in the simplex $\Delta(K)$ and starting at p.
Maximizing on the corresponding set gives, by Jensen's inequality, $v(p) = \mathtt{Cav}\, u(p)$.
b) A parallel approach in the dual game is very effective, Laraki (2000a).
The recursive formula for the dual game (Proposition 3.23) leads to consider w_n as the nth upper discretization of a differential game on $[0, 1]$ with dynamic $z(\theta)$ satisfying:

$$\frac{dz}{d\theta} = s_\theta G t_\theta, \quad z(0) = -x$$

and terminal payoff $\max_k z^k(1)$. (Compare with Theorem B6).
One then shows that the game starting at time θ from state z has a value $\varphi(\theta, z)$ being the only viscosity solution, uniformly continuous in z uniformly in θ, of:

$$\frac{\partial \varphi}{\partial \theta} + u(\nabla\varphi) = 0, \quad \varphi(1,z) = \max_k z^k \quad (H).$$

One obtains $\lim_{n\to\infty} w_n(x) = w(x) = \varphi(0,-x)$. Thanks to the time homogeneity property,

$$\varphi(\theta,x) = (1-\theta)\varphi(0,\frac{x}{1-\theta}),$$

w is a viscosity solution of

$$f(x) - \langle x, \nabla f(x)\rangle - u(-\nabla f(x)) = 0, \quad \lim_{\alpha\to 0} \alpha f(\frac{x}{\alpha}) = max_k - x^k$$

which is equation (E) in section 5.2, but with a recession limit condition. One can identify the solution of (H), written with $\psi(\theta,x) = \varphi(1-\theta,x)$ as :

$$\frac{\partial \psi}{\partial \theta} + L(\nabla\psi) = 0 \quad \psi(0,x) = b(x)$$

with L continuous, b uniformly Lipschitz and convex, using Hopf's formula:

$$\psi(\theta,x) = \sup_{p\in \mathbb{R}^K} \inf_{q\in \mathbb{R}^K} \{b(q) + \langle p, x-q\rangle - \theta L(p)\}$$

which gives here

$$\psi(\theta,x) = \sup_{p\in \mathbb{R}^K} \inf_{q\in \mathbb{R}^K} \{\max_k q^k + \langle p, x-q\rangle + \theta u(p)\}$$

and finally $w(x) = \psi(1,-x) = \sup_{p\in\Delta(K)}\{u(p) - \langle p,x\rangle\} = \Lambda_s(u)(x)$.
Going back to the primal game one obtains

$$\lim_{n\to\infty} v_n(p) = \Lambda_i \circ \Lambda_s(u)(p) = \mathsf{Cav}\ u(p).$$

3.7.3 Speed of convergence

Concerning the speed of convergence of v_n (resp. v_λ) to Cav u in Theorems 3.16 and 3.19 much more precise results are available:
First, in an example due to Zamir (1971-72) with $K = 2$, one has Cav $u \equiv 0$ and $v_n(p) \geq \dfrac{p(1-p)}{\sqrt{n}}$ which shows that the order of convergence of $\frac{1}{\sqrt{n}}$ is the best one.
Mertens and Zamir (1976b) then started the study of the first term in the asymptotic expansion:

$$\delta_n = \sqrt{n}(v_n - \mathsf{Cav}\ u)$$

and proved, for a class of games including Zamir's example, the convergence of $\delta_n(p)$ to $\varphi(p) = \frac{1}{\sqrt{2\pi}} e^{-x_p^2/2}$, the evaluation of the density of the Normal Law at its p-quantile x_p defined by:

$$p = \int_{-\infty}^{x_p} \frac{e^{-x^2/2}}{\sqrt{2\pi}} dx$$

Mertens and Zamir (1977b) actually showed that this quantity is equal to the maximal variation of a martingale $\mathbf{q}$ in $[0,1]$ with expectation p:

$$\lim_{n\to\infty} \sup_{\mathbf{q}} \frac{\mathbf{V}_n^1(\mathbf{q})}{\sqrt{n}} = \varphi(p).$$

Thus in the above game the optimal strategy of Player 1 generates the martingale of posteriors that has the largest L^1 variation.
The appearance of the Normal law was later on explained through an analysis describing explicitely optimal strategies for Zamir's game by Heuer (1992a) and more generally through the dual game by De Meyer (1996a, 1996b). This leads to a new approach of the compact game in continuous time where the random past is represented by a Brownian motion, De Meyer (1999).
Finally a direct proof for the evaluation of the maximal variation of a martingale has been obtained by De Meyer (1998) with similar methods of stochastic calculus.

3.7.4 Signals

The model introduced in Section 1 can be generalized by adding signals after each stage.

The simplest framework (Level 1) corresponds to a signalling scheme ℓ for Player 2. ℓ is a map from I to some signal space B and after each stage m, Player 2 is told $\ell(i_m)$. The consequence is that Player 1 may use his information, namely play differently as a function of k, without revealing it. Thus the two notions of not revealing and not using the information are no longer equivalent. However it follows from the analysis in Section 2 that the crucial role is still played by the posterior p_m, hence by the information transmitted. One thus defines $NR(p) = \{s \in S^K \colon p^k p^{k'} > 0 \Rightarrow \ell(s^k) = \ell(s^{k'}),\ \forall k, k'\}$ (where $\ell(s^k)$ is the usual linear extension) which corresponds to the vectors of mixed moves inducing, for each k, the same distribution on signals. This is clearly a convex polyhedron hence the non-revealing game on $NR(p) \times T$ with payoff $\sum_{k \in K} p^k s^k G^k t$ has still a value u. All the previous analysis goes trough. The extension to random signals is clear.

A more general case (Level 2) is where ℓ is a map from $I \times J$ to B (or even $\Delta(B)$). Now the information that Player 2 obtains is also function of his own move. One has $NR(p) = \{s \in S^K \colon p^k p^{k'} > 0 \Rightarrow \ell(s^k, j) = \ell(s^{k'}, j),\ \forall j \in J, \forall k, k'\}$. The non-revealing game is defined as above and the convergence of v_n to $\mathsf{Cav}\, u$ still holds. However, due to the cost of information for Player 2 who may have to use specific moves to obtain revealing signals, the speed of convergence is lower, namely $\sqrt[3]{n}$ (MSZ, Chapter VI) and this is the best bound (Zamir, 1973a).

Note that in this model Player 1 may not know the information of Player 2, hence the recursive formula 3.20 does not hold.

Level 3 corresponds to an even more general model given by a signalling map ℓ from $K \times I \times J$ to $\Delta(B)$. This covers cases where Player 1 has to use his information to play non revealing and also where Player 1 cannot prevent Player 2 to learn about the true state. The set $NR(p)$ can now be empty for some values of p (u is then $-\infty$) but u is finite on the extreme points of $\Delta(K)$, hence $\mathtt{Cav}\ u$ is well defined. In this framework Theorem 3.32 still holds: v_∞ exists and equals $\mathtt{Cav}\ u$ (Aumann and Maschler, 1995, Chapter 5, first version 1968).
The construction in this case of the "approachability strategy" of the uninformed player (Corollary 3.34) was done by Kohlberg (1975): since the moves are not necesserily observed, the sequence of past payoffs is unknown and Blackwell's procedure does not apply immediately. The proof is much more involved: one of the basic ideas is to construct an evaluation of the payoff based on a statistics of the past signals.
A recent result of Mertens (1998) gives a bound of $O(\frac{Log\ n}{n})^{1/3}$ on the speed of convergence of v_n to its limit.
Remark finally that in all cases above the value of the infinite game exists and is independent of the signalling scheme for Player 1.

3.8 Exercises

3.8.1

Check the assertions in the proof of Proposition 3.27.
(Consider a Taylor expansion of f_n at $\frac{n+1}{n}x$ in $\Psi_{n+1} f_n(x)$, see De Meyer and Rosenberg (1999), Lemmas 2,3 and 4).

3.8.2 Study of v_λ (Mayberry, 1967)

a) Use Proposition 3.20 to obtain in the framework of Example 1, Chapter 1 that the discounted value satisfies:

$$v_\lambda(p) = \max_{s^1,s^2}\{\lambda \min_k p^k s^k + (1-\lambda)E(v_\lambda(\tilde{p}))\}$$

and check that v_λ is symmetric around 1/2.
b) Deduce that the maximum is obtained for $p^1 s^1 = p^2 s^2 = \alpha$ and that for $p^1 \in [0, 1/2]$, one has:

$$v_\lambda(p) = \max_{0 \le \alpha \le p^1}\{\lambda\alpha + (1-\lambda)(p^1 v_\lambda(\frac{(p^1-\alpha)}{p^1}) + p^2 v_\lambda(\frac{\alpha}{p^2})\}$$

c) Show that for $\lambda \in [2/3, 1]$ the previous equation gives

$$v_\lambda(p) = \lambda p^1 + (1-\lambda)p^2 v_\lambda(\frac{p^1}{p^2}) \qquad (**)$$

d) Observe that v_λ at a rational point can be express as a function of the values at rational points with smaller denominator. For example prove inductively that:

$$v_\lambda(1/2) = \frac{\lambda}{2}, \qquad v_\lambda(1/3) = \frac{\lambda(2-\lambda)}{3}, \ldots$$

f) Differentiate equation(**) to obtain, with $p = p^1 = 1 - p^2$:

$$\frac{dv_\lambda}{dp}(p) = (1-\lambda)(1 - \frac{p}{1-p})\frac{dv_\lambda}{dp}(\frac{p}{1-p}) - (1-\lambda)v_\lambda(\frac{p}{1-p})$$

from which it follows that left and right derivatives at $p = 1/2$ exist and differ. Deduce from d) that the derivatives at rational points will be express as functions of $\frac{dv_\lambda}{dp}(1/2)$.
g) Conclude that v_λ has discontinuous derivative at each rational point.

3.8.3 Subadditivity

Prove that

$$v_{kn+m} \leq kv_n + v_m$$

and that this implies the existence of $\lim v_n$.

3.9 Notes

The model and all the results of this Section are, unless indicated below, due to Aumann and Maschler. They were developed in a series of reports prepared by Mathematica for the United States Arms Control and Disarmament Agency in 1966, 1967, 1968, see Aumann and Maschler (1995).
The first approach of the discounted game is in Mayberry (1967).
The argument in Proposition 3.7 is quite general and goes back to Gleason.
Proposition 3.9 comes from MSZ (V, Th 3.1). Proposition 3.11 is in Lehrer and Yariv (1998) and Proposition 3.22 is in Sorin (1979).
Proposition 3.23 and the next Corollaries were obtained by De Meyer (1996a) and generalized by Rosenberg (1998).
The content of Section 5.2 follows De Meyer and Rosenberg (1998).
The result of Section 5.3, due to Mertens, appeared in MSZ (V, Ex. 6.5).

A much more general analysis can be found in MSZ, Chapter V.

4 Repeated games with lack of information on both sides

4.1 Presentation

In this chapter again the model of game with lack of information on both sides is introduced in the simplest framework: standard signalling and independent case. The description is thus very similar to the one given in Chapter 3, Section 3.1. (See Section 7 for extensions.)
K and L are finite sets and $\{G^{k\ell}; k\in K, \ell\in L\}$ denotes a family of $I\times J$ real matrices. Let $\|G\| = \max_{ijk\ell} |G^{k\ell}_{ij}|$.
For each couple $(p,q)\in\Delta(K)\times\Delta(L)$, the **game form** $\mathcal{G}(p,q)$ is as follows:
- at stage one, k (resp. ℓ) is chosen according to the probability p (resp. q) and is announced to Player 1 (resp. Player 2) only.
- at stage m, knowing the past history $h_m = (i_1, j_1, \ldots, i_{m-1}, j_{m-1})$, Player 1 (resp. Player 2) chooses i_m in I (resp. j_m in J) and the new history $h_{m+1} = (h_m, i_m, j_m)$ is told to both.
The previous description is public knowledge.
$H_m = (I\times J)^{m-1}$ is the set of histories at stage m and $H = \cup_{m\geq 1} H_m$ the set of histories.
A strategy σ in Σ for Player 1 (resp. τ in $\mathcal{T}$ for Player 2) is a map from $K\times H$ to $S = \Delta(I)$ (resp. $L\times H$ to $T = \Delta(J)$).
The expectation on the set $K\times L\times(I\times J)^\infty$ of plays, endowed with theusual cylindar σ-field, is with respect to the probability induced by (p,q,σ,τ), and $\gamma^{p,q}_m(\sigma,\tau)$ denotes the expected payoff at stage m: $\gamma^{p,q}_m(\sigma,\tau) = E^{p,q}_{\sigma,\tau}(g_m)$ with g_m being the random variable $G^{kl}_{i_m j_m}$.
As in Chapter 3, Section 3.1, one introduces the n stage game $G_n(p,q)$ and the discounted game $G_\lambda(p,q)$ with values $v_n(p,q)$ and $v_\lambda(p,q)$ respectively. The definitions of the maxmin $\underline{V}(p,q)$ and of the minmax $\overline{V}(p,q)$ of the infinite game $G_\infty(p,q)$ are also similar.

4.2 Basic properties

We define first a quantity that will play a crucial role in the analysis.

Definition
The **informational content** of a strategy $\sigma \in \Sigma$ is:

$$\mathcal{I}(\sigma) = \sup_{\nu : H \to J} E_{\sigma,\nu}\left(\sum_k \sum_{m=1}^{\infty} (p_{m+1}^k - p_m^k)^2\right).$$

This measures the information that can be extracted from this strategy during a play of $\mathcal{G}(p,q)$.
Note that the supremum is taken on the set of "pure non revealing" strategies of Player 2 (or equivalently on the plays compatible with σ), but one has:

Lemma 4.1

$$\mathcal{I}(\sigma) = \sup_{\tau \in \mathcal{T}} E_{\sigma,\tau}\left(\sum_k \sum_{m=1}^{\infty} (p_{m+1}^k - p_m^k)^2\right).$$

Proof
Remark that $E_{\sigma,\tau}$ can be written $\sum_\ell q^\ell E_{\sigma,\tau^\ell}$ since the integrand is independent of ℓ. Then it is enough to consider extreme points. ∎

Given $\varepsilon > 0$ and σ, let $\nu^* : H \to J$ and N such that:

$$\mathcal{I}(\sigma) - \varepsilon \leq E_{\sigma,\nu^*}\left(\sum_k \sum_{m=1}^{N} (p_{m+1}^k - p_m^k\right)^2)$$

then the following holds:

Lemma 4.2
Given σ and ε, for any strategy τ that coincides with ν^* up to stage N and any $n \geq N$:

$$E_{\sigma,\tau}(\|p_{n+1} - p_{N+1}\|_1) \leq \sqrt{K\varepsilon}.$$

Proof
By definition of ν^* and N, for any $n \geq N$:

$$E_{\sigma,\tau}\left(\sum_k (p_{n+1}^k - p_{N+1}^k)^2\right) = E_{\sigma,\tau}\left(\sum_k \sum_{m=N+1}^{n} (p_{m+1}^k - p_m^k)^2\right) \leq \varepsilon$$

so that the result follows from Cauchy-Schwartz inequality.

∎

In words, since ν^* extracts up to stage N the maximal amount of information from σ, σ is essentially non-revealing after stage N.

As in the previous chapter the **non-revealing game** is the game $D(p,q)$ played on $NR^1(p) \times NR^2(q)$ where:

$$NR^1(p) = \{\{s^k\}_{k \in K} : p^k p^{k'} > 0 \Rightarrow s^k = s^{k'}\}$$

is the set of **non-revealing strategies** for Player 1 (and similarly $NR^2(q)$ for Player 2). Its value $u(p, q)$ is also the value of the game on $S \times T$ with payoff $\sum_{k\in K, \ell\in L} p^k q^\ell G^{k,\ell}$. $D(p, q)$ is thus the game where the players do no transmit any information (or equivalently here, are not using their information).
In contrast with the case of lack of information on one side, the analysis of the infinite game $G_\infty(p, q)$ is somehow easier than the one of the compact game and will be covered first.

Notation
Recall that $\mathtt{Cav}$ denotes the concavification operator . Similarly $\mathtt{Vex}$ denotes the convexification operator so that $\mathtt{Vex}(f) = -\mathtt{Cav}(-f)$ (See appendix A.8). For f defined on $\Delta(K)\times\Delta(L)$, we denote by $\mathtt{Cav}_p f(p', q')$ the value at p' of the concavification of the function $f(., q')$ on $\Delta(K)$. Similarly $\mathtt{Vex}_q$ is the convexification operator with respect to q on $\Delta(L)$, p being fixed.

4.3 Infinite game

In this section the existence and characterization of the minmax $\overline{V}(p, q)$ and of the maxmin $\underline{V}(p, q)$ will be obtained. (Recall the definitions in Chapter 3, Section 3.1).

Proposition 4.3
Player I can guarantee $\mathtt{Vex}_q u$.

Proof
Let Player 1 forget his information. The situation is now similar to a game with lack of information on one side where Player 2 is informed, and defined by: the set L, the probability $q\in\Delta(L)$ and the family of matrices $B^\ell(p) = \sum_{k\in K} p^k G^{k\ell}$. The value of the corresponding non revealing game is $\tilde{u}(q) = \mathtt{val}_{S\times NR^2(q)} \sum_{\ell\in L} q^\ell s B^\ell(p) t^\ell = \mathtt{val}_{S\times T} \sum_{\ell\in L} q^\ell B^\ell(p) = u(p, q)$. Applying the results of Chapter 3, Section 3.6, we obtain from Proposition 3.31 (or more precisely its dual because the uninformed player maximizes) that Player 1 can guarantee $\mathtt{Vex}\ \tilde{u} = \mathtt{Vex}_q u$.

∎

Corollary 4.4
Player 1 can guarantee $\mathtt{Cav}_p \mathtt{Vex}_q u$.

Proof
The proof follows from the previous result and Lemma 3.30 which extends directly to this case, with the same argument (splitting property). ∎

Proposition 4.5
Player 2 can defend $\mathtt{Cav}_p \mathtt{Vex}_q u$.

Proof
Given σ, Player 2 follows up to stage N a strategy ν^* exhausting, up to ε the informational content $\mathcal{I}(\sigma)$. From stage $N+1$ on, he plays an optimal strategy in the game with lack of information on one side starting at state (p_{N+1}, q) and where he is informed. Thus the game is characterized by the set L, the probability $q \in \Delta(L)$ and the family of matrices $B^\ell(p_{N+1}) = \sum_{k\in K} p^k_{N+1} G^{k\ell}$. Note that since Player 2 is using a non-revealing strategy up to stage N, the state variable on L at stage $N+1$ is still q.
On the other hand if $\rho_m(\sigma,\tau)(h_m) = \sum_{k\in K,\ell\in L} p^k_m(h_m) q^\ell_m(h_m)\sigma^k(h_m)G^{k\ell}\tau^\ell(h_m)$ denotes the payoff in the local game at h_m, one has, using Proposition 3.2 and Lemma 3.13:

$$\rho_m(\sigma,\tau)(h_m) \le \sum_{k,\ell} p^k_m(h_m) q^\ell_m(h_m)\bar{\sigma}^k(h_m)G^{k\ell}\tau^\ell(h_m)$$
$$+\|G\|E(\|p_{m+1}-p_m\|_1|h_m)$$

Hence, letting $\zeta_m(\sigma,\tau)(h_m) = \sum_\ell q^\ell_m(h_m)\bar{\sigma}^k(h_m)(\sum_k p^k_{N+1}(h_m)G^{k\ell})\tau^\ell(h_m)$, one obtains:

$$\rho_m(\sigma,\tau)(h_m) \le \zeta_m(\sigma,\tau)(h_m)+\|G\|(E(\|p_{m+1}-p_m\|_1|h_m)+\|p_m(h_m)-p_{N+1}\|_1).$$

Observe that $\zeta_m(\sigma,\tau)(h_m)$ is the payoff given h_m induced by τ and a nonrevealing strategy of Player 1 in the local game at (p_{N+1}, q_m). The choice of τ and Proposition 3.28 implies that for M large enough:

$$E\left(\sum_{N+1}^{N+M} \zeta_m(\sigma,\tau)(h_m)|\mathcal{H}_{N+1}\right) \le M(\mathtt{Vex}_q u(p_{N+1},q) \mid c).$$

Recall that $E(\|p_{m+1}-p_m\|_1|h_m) = LV(p,\sigma)(h_m)$. Using Lemma 4.2 it follows that:

$$E\left(\sum_{m=1}^{M+N} \rho_m(\sigma,\tau)(h_m)\right) \le (M+N)E(\mathtt{Vex}_q u(p_{N+1},q)+\varepsilon)$$
$$+\|G\|(2N+M\sqrt{K\varepsilon}+\sqrt{M}).$$

It remains to majorize $\mathtt{Vex}_q u$ by $\mathtt{Cav}_p\mathtt{Vex}_q u$ and to use Jensen's inequality to obtain:
$\forall\eta, \exists N_0$ such that for all $n \ge N_0$:

$$\bar{\gamma}_n(\sigma,\tau) \le \mathtt{Cav}_p\mathtt{Vex}_q u(p,q)+\eta.$$

∎

The proofs of the two previous results are somehow dual. Player 1 uses his information at stage one to perform the splitting that realizes the concavification. From then on he plays non-revealing in the game where Player 2 is informed.

The reply of Player 2 shows that Player 1 cannot do better: after finitely many stages, Player 1 will be essentially playing non-revealing, hence Player 2 can obtain from this time on, the maxmin (here the value) of the game where he is the only informed player.
Corollary 4.4 and Proposition 4.5 thus imply:

Theorem 4.6
The maxmin and the minmax of $G_\infty(p,q)$ exist and are respectively:

$$\underline{V}(p,q) = \mathtt{Cav}_p\mathtt{Vex}_q u(p,q)$$

$$\overline{V}(p,q) = \mathtt{Vex}_q\mathtt{Cav}_p u(p,q).$$

It follows that $v_\infty(p,q)$ exists iff the two operators commute for u at (p,q):

$$\mathtt{Cav}_p\mathtt{Vex}_q u(p,q) = \mathtt{Vex}_q\mathtt{Cav}_p u(p,q).$$

The first examples of games for which this equality does not hold are due to Aumann and Maschler (1995, first version 1967). A simple piece wise bilinear function f on the square $\Delta(K)\times\Delta(L) = [0,1]^2$, for which $\mathtt{Cav}_p\mathtt{Vex}_q f$ and $\mathtt{Vex}_q\mathtt{Cav}_p f$ differ is given by the following values (extended by bilinearity):

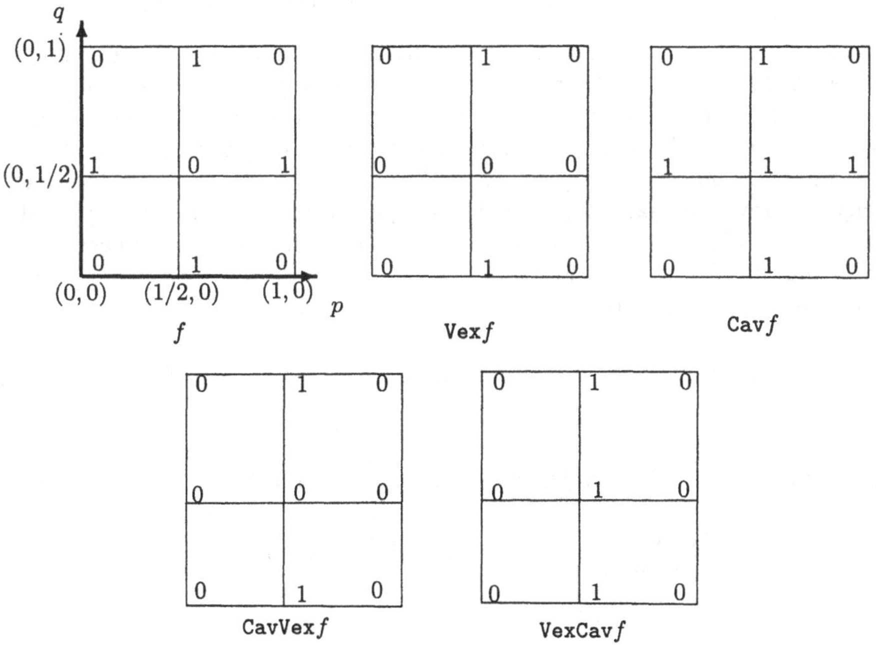

Fig. 4.1.

4.4 Asymptotic approach

The existence of a value for finitely repeated or discounted games is clear, as in Chapter 3. We consider here the asymptotic behavior of these families.

4.4.1 Finite game

$G_n(p,q)$ is a finite game, hence $v_n(p,q)$ exists and the properties of Chapter 2 apply. Let $\underline{w} = \liminf_{n\to\infty} v_n$ and $\overline{w} = \limsup_{n\to\infty} v_n$. Then $\underline{w}$ is concave w.r.t. p, Lipschitz with constant $\|G\|$ and $d_n = (\underline{w} - v_n)^+$ converges uniformly to 0 on $\Delta(K)\times\Delta(L)$.

Proposition 4.7

$$\underline{w}(p,q) \geq \mathtt{Cav}_p\mathtt{Vex}_q \max\{u(p,q), \underline{w}(p,q)\},\ \forall p,q\in\Delta(K)\times\Delta(L)$$

$$\overline{w}(p,q) \leq \mathtt{Vex}_q\mathtt{Cav}_p \min\{u(p,q), \overline{w}(p,q)\},\ \forall p,q\in\Delta(K)\times\Delta(L).$$

Proof
We prove the first inequality.
Let us start by showing that:

$$\forall\varepsilon>0, \exists N, n\geq N \Rightarrow v_n(p,q) \geq \mathtt{Vex}_q \max\{u(p,q), \underline{w}(p,q)\} - \varepsilon.$$

Using the minmax theorem it is thus enough to prove that for any $\tau\in\mathcal{T}$ there exists σ with $\overline{\gamma}_n^{pq}(\sigma,\tau) \geq \mathtt{Vex}_q \max\{u(p,q), \underline{w}(p,q)\} - \varepsilon$. Knowing τ, Player 1 can compute q_m, stage after stage. σ is defined inductively in the local game after each history h_m using the stopping time $\theta = \min\{m; u(p_m,q_m) < \underline{w}(p_m,q_m)\}$. Explicitly σ_m is optimal in the one stage non-revealing game $D(p_m,q_m)$ as long as $m<\theta$. After h_θ, σ is optimal in the game $G_{n-\theta+1}(p_\theta,q_\theta)$, which corresponds to the remaining stages. Note that by definition σ is non revealing before θ, hence $p_\theta = p$. Using Proposition 3.14 we thus obtain:

$$\begin{aligned}\gamma_n^{pq}(\sigma,\tau) \geq E\{(&\sum\nolimits_{m=1}^{\theta-1}(u(p_m,q_m) - \|G\|\|q_{m+1}-q_m\|_1)\\ &+(n-\theta+1)v_{n-\theta+1}(p,q_\theta)\}.\end{aligned}$$

Hence by Lemma 3.5 (as in the proof of proposition 3.15):

$$\begin{aligned}\gamma_n^{pq}(\sigma,\tau) \geq E&\left\{\sum\nolimits_{m=1}^{\theta-1}u(p,q_m) + (n-\theta+1)(\underline{w}(p,q_\theta) - \|d_{n-\theta+1}\|)\right\}\\ &-\#L\|G\|\sqrt{n}.\end{aligned}$$

Let us first evaluate $B = E\{\sum_{m=1}^{\theta-1}u(p_m,q_m) + (n-\theta+1)\underline{w}(p,q_\theta)\}$. By the choice of θ, $u(p,q_m) = \max\{u,\underline{w}\}(p,q_m)$ and $\underline{w}(p,q_\theta) = \max\{u,\underline{w}\}(p,q_\theta)$. So that:

$$B \geq E\left\{\sum_{m=1}^{\theta-1} \max\{u, \underline{w}\}(p, q_m) + (n - \theta + 1)\max\{u, \underline{w}\}(p, q_\theta)\right\}$$

and a fortiori the inequality holds replacing $\max\{u, \underline{w}\}$ by $\mathtt{Vex}_q \max\{u, \underline{w}\}$. Hence Jensen's inequality gives:

$$B \geq n\ \mathtt{Vex}_q \max\{u, \underline{w}\}(p, E\{\frac{1}{n}\sum_{m=1}^{\theta-1} q_m + (n - \theta + 1)q_\theta\})$$

thus, $\{q_m\}$ being a martingale:

$$B \geq n\ \mathtt{Vex}_q \max\{u, \underline{w}\}(p, q).$$

Now let N such that $n \geq N$ implies $d_N \leq \varepsilon/3$, then for $n \geq N_1 = \max(N, 6N\|A\|/\varepsilon)$ one obtains:

$$\frac{1}{n}E\{(n - \theta + 1)\|d_{n-\theta+1}\|\} \leq \frac{N}{n}2\|G\| + \frac{\varepsilon}{3} \leq \frac{2}{3}.$$

So that, for $n \geq N_2 = \max(N_1, (3\|A\|\#L\ /\varepsilon)^2)$ the following holds:

$$\gamma_n^{pq}(\sigma, \tau) \geq \mathtt{Vex}_q \max\{u, \underline{w}\}(p, q) - \varepsilon$$

thus:

$$\underline{w} \geq \mathtt{Vex}_q \max\{u, \underline{w}\}.$$

Since $\underline{w}$ is concave in p, one deduces:

$$\underline{w} \geq \mathtt{Cav}_p\mathtt{Vex}_q \max\{u, \underline{w}\}$$

and similarly

$$\overline{w} \leq \mathtt{Vex}_q\mathtt{Cav}_p \min\{u, \overline{w}\}.$$

■

We now majorize the functions satisfying the same inequality than $\overline{w}$.

Proposition 4.8
Let f a real function defined on $\Delta(K)\times\Delta(L)$, with:

$$f(p, q) \leq \mathtt{Vex}_q\mathtt{Cav}_p \min\{u(p, q), f(p, q)\},$$

then:

$$f(p, q) \leq v_n(p, q) + \frac{\|G\|}{\sqrt{n}}\sum_{\ell\in L}\sqrt{q^\ell(1 - q^\ell)},$$

hence, in particular:

$$f \leq \underline{w}.$$

Proof
Let f be maximal among the functions satisfying the above inequality. One

then has (see Exercise 8.1): $f(p,q) = \mathtt{Vex}_q\mathtt{Cav}_p \min\{u(p,q), f(p,q)\}$, hence assume f convex in q.
Moreover Caratheodory's Theorem implies:
$\forall \varepsilon_0 > 0, \forall (p,q) \in \Delta(K) \times \Delta(L)$, there exists a family $p(r) \in \Delta(K)$, with $r \in K$ and $\lambda \in \Delta(K)$ adapted to p at q, i.e. such that:

$$\sum\nolimits_{r\in K} \lambda_r p(r) = p \quad \text{and} \quad f(p,q) - \varepsilon_0 \leq \sum\nolimits_{r\in K} \lambda_r \min\{u,f\}(p(r),q).$$

Given a strategy τ of Player 2, we define a strategy σ of player 1 as follows:
- as long as $u(p_m,q_m) \geq f(p_m,q_m)$, play optimally in the non-revealing local game $D(p_m,q_m)$,
- as soon as $u(p_m,q_m) < f(p_m,q_m)$, use the splitting property (Proposition 2.3) to generate the family $\{p_m(r)\}$, adapted to p_m at q_m as above, by playing optimally in the non-revealing game at $(p_m(r),q_m)$.
One thus obtain, in the first case, using Proposition 3.14:

$$\rho_m(\sigma,\tau)(h_m) \geq u(p_m(h_m),q_m(h_m)) - \|G\| LV(\tau,q)(h_m)$$

and in the second case:

$$\rho_m(\sigma,\tau)(h_m) \geq E(u(p_{m+1},q_m)|h_m) - \|G\| LV(\tau,q)(h_m).$$

But one has:

$$E(u(p_{m+1},q_m)|h_m) = \sum \lambda_r u(p_m(r)(h_m),q_m(h_m)) \geq f(p_m(h_m),q_m(h_m)) - \varepsilon_0.$$

Hence in both cases, by the definition of τ:

$$\rho_m(\sigma,\tau)(h_m) \geq f(p_m(h_m),q_m(h_m)) - \varepsilon_0 - \|G\| LV(\tau,q)(h_m).$$

The choice of $p_m(r)$ also implies:

$$E(f(p_{m+1},q_m)|h_m) \geq f(p_m(h_m),q_m(h_m)) - \varepsilon_0$$

and f is convex w.r.t. q so that, using the conditional independence of p_{m+1} and q_{m+1}, given h_m:

$$E(f(p_{m+1},q_{m+1})|h_m) \geq f(p_m(h_m),q_m(h_m)) - \varepsilon_0$$

hence by induction:

$$E(f(p_m,q_m)) \geq f(p,q) - (m-1)\varepsilon_0$$

so that, using Proposition 3.15, $\forall n$, $\forall \varepsilon = \varepsilon_0 n$:

$$\overline{\gamma}_n^{pq}(\sigma,\tau) \geq f(p,q) - \varepsilon - \frac{\|G\|}{\sqrt{n}} \sum\nolimits_{\ell\in L} \sqrt{q^\ell(1-q^\ell)}$$

which gives the result. ∎

Note that this proof corresponds to both aspects that appeared in Chapter 3: Player 1 can obtain Cav u trough a splitting and Player 2 by playing non-revealing.

Proposition 4.9

$\lim_{n\to\infty} v_n = w$ exists and satisfies:

$$-\frac{\|G\|}{\sqrt{n}}\sum\nolimits_{\ell}\sqrt{q^{\ell}(1-q^{\ell})} \;\le\; (v_n - w)(p,q) \;\le\; \frac{\|G\|}{\sqrt{n}}\sum\nolimits_{k}\sqrt{p^{k}(1-p^{k})}.$$

Proof

Proposition 4.7 shows that $\overline{w}$ satisfies the conditions of Proposition 4.8. Hence $\overline{w} = \underline{w}$ and $\lim_{n\to\infty} v_n$ exists. The bounds follow then from Proposition 4.8.

■

4.4.2 Discounted game

The existence of $v_\lambda(p,q)$ is as in Chapter 3, Section 3.1. The proof that the limit of the discounted values exists is completely similar to the previous one. Explicitly let $\underline{v} = \liminf_{\lambda\to 0} v_\lambda$ and $\overline{v} = \limsup_{\lambda\to 0} v_\lambda$. One obtains, using Proposition 3.18:

Proposition 4.10

$$\underline{v} \ge \mathtt{Cav}_p\mathtt{Vex}_q \max\{u, \underline{v}\}$$

$$\overline{v} \;\le\; \mathtt{Vex}_q\mathtt{Cav}_p \min\{u, \overline{v}\}.$$

Proof

The structure of the proof is as in Proposition 4.7.

Player 1 uses σ optimal in the local non revealing game until stage $\theta = \min\{m;\ u(p_m, q_m) < \underline{v}(p_m, q_m)\}$. From this stage on, he plays optimally in $v_\lambda(p_\theta, q_\theta)$.

■

The analog of Proposition 4.8 is:

Proposition 4.11

Let f a real function defined on $\Delta(K)\times\Delta(L)$, with:

$$f(p,q) \;\le\; \mathtt{Vex}_q\mathtt{Cav}_p \min\{u(p,q), f(p,q)\}.$$

Then:

$$f(p,q) \;\le\; v_\lambda(p,q) + \|G\|\sum\nolimits_{\ell}\sqrt{q^{\ell}(1-q^{\ell})}\frac{\sqrt{\lambda}}{\sqrt{2-\lambda}}$$

hence:

$$f \;\le\; \underline{v}.$$

Proof
The main steps are like in Proposition 4.8.
The last line is now:

$$\gamma_\lambda^{pq}(\sigma,\tau) \geq f(p,q) - \frac{\varepsilon_0}{\lambda} - \|G\|\sum\nolimits_\ell \sqrt{q^\ell(1-q^\ell)}\frac{\sqrt{\lambda}}{\sqrt{2-\lambda}}.$$

∎

Similarly one has:

Proposition 4.12
$\lim_{\lambda\to 0} v_\lambda = \nu$ *exists and satisfies:*

$$-\|G\|\frac{\sqrt{\lambda}}{\sqrt{2-\lambda}}\sum\nolimits_\ell \sqrt{q^\ell(1-q^\ell)} \;\leq\; (v_\lambda - \nu)(p,q) \;\leq\; \|G\|\frac{\sqrt{\lambda}}{\sqrt{2-\lambda}}\sum\nolimits_k \sqrt{p^k(1-p^k)}$$

4.5 The functional equation

We prove here that the functional equation satisfied by w and ν has a unique solution, in particular $\lim_{n\to\infty} v_n = \lim_{\lambda\to 0} v_\lambda$. (An alternative direct proof of existence and equality of these limits can be found in Subsection 6.2).

4.5.1 Basic properties

We introduce the following inequalities:

$$(a) \qquad f \;\geq\; \mathtt{Cav}_p\mathtt{Vex}_q \max\{u, f\}$$

$$(b) \qquad f \;\leq\; \mathtt{Vex}_q\mathtt{Cav}_p \min\{u, f\}.$$

Proposition 4.13
w is the smallest solution of (a) and the largest solution of (b).
In particular w is the only solution of both. Same results hold for ν.

Proof
By Proposition 4.7, w satisfies (a) and (b). By Proposition 4.8, any solution f of (b) satisfies $f \leq \underline{w} = w$. A dual property holds for (a). (Same arguments apply for ν). ∎

From now on we write v for both w and ν. f defined on $\Delta(K)\times\Delta(L)$ is a **saddle function** if it is concave in p and convex in q.
Let us introduce two implicit functional equations:

$$(a') \qquad f = \mathtt{Vex}_q \max\{u, f\}$$

$$(b') \qquad f = \mathtt{Cav}_p \min\{u, f\}.$$

Proposition 4.14
v is the only solution of (a') and (b').

Proof
If f satisfies (a') and (b'), f is a saddle function. Hence it satisfies both (a) and (b), thus equals v by Proposition 4.13.
On the other hand v satisfies (a) hence $v \geq \mathtt{Vex}_q \max\{u, v\}$. But v is convex in q (like $\overline{w}$) and thus $v \leq \max\{u, v\}$ implies the reverse inequality $v \leq \mathtt{Vex}_q \max\{u, v\}$. ∎

4.5.2 An iterative approach

Let $x_0 \equiv -\infty$ and define inductively x_{n+1} on $\Delta(K)\times\Delta(L)$ by :

$$x_{n+1} = \mathtt{Cav}_p\mathtt{Vex}_q \max\{u, x_n\}$$

u is $\|G\|$ Lipschitz on a product of simplices hence the increasing sequence $\{x_n\}$ converges uniformly to some $\|G\|$ Lipschitz function x satisfying:

$$x = \mathtt{Cav}_p\mathtt{Vex}_q \max\{u, x\}.$$

One introduces similarly on $\Delta(K)\times\Delta(L)$ $y_0 = -\infty$, then:

$$y_{n+1} = \mathtt{Vex}_q\mathtt{Cav}_p \min\{u, y_n\}$$

and the sequence converge to some y with:

$$y = \mathtt{Vex}_q\mathtt{Cav}_p \min\{u, y\}.$$

Proposition 4.15

$$\underline{w} \geq x, \quad \overline{w} \leq y.$$

Proof
The proof of Proposition 4.7 shows that if $\underline{w} \geq x_n$ then $\underline{w} \geq x_{n+1}$ and the result follows by induction.

The next result is in the spirit of the maximum principle.

Proposition 4.16

$$x \geq y.$$

Proof
By contradiction, assume the function $(y - x)$ to be positive at some point of $\Delta(K)\times\Delta(L)$. Then the subset where $(y - x)$ is maximal and equal to

some $\delta > 0$, is non empty and compact. Denote by (p^*, q^*) an extreme point of its convex hull. We claim that one of the operators $\mathtt{Vex}_q \max\{u, x\}$ or $\mathtt{Cav}_p \min\{u, y\}$ is non trivial at that point. Otherwise:

$$x(p^*, q^*) = \mathtt{Cav}_p\mathtt{Vex}_q \max\{u, x\}(p^*, q^*) \geq$$
$$\mathtt{Vex}_q \max\{u, x\}(p^*, q^*) = \max\{u, x\}(p^*, q^*)$$

and

$$y(p^*, q^*) = \mathtt{Vex}_q\mathtt{Cav}_p \min\{u, y\}(p^*, q^*) \leq$$
$$\mathtt{Cav}_p \min\{u, y\}(p^*, q^*) = \min\{u, y\}(p^*, q^*)$$

would contradict: $x < y$ at (p^*, q^*).
Assume thus a non trivial splitting:

$$\mathtt{Vex}_q \max\{u, x\}(p^*, q^*) = \sum_{r \in L} \mu^r \max\{u, x\}(p^*, q^*(r))$$

with μ in $\Delta(L)$, $q^*(r)$ in $\Delta(L)$ and $\sum_{r \in L} \mu^r q^*(r) = q$. But then:

$$\begin{aligned} x(p^*, q^*) &= \mathtt{Cav}_p\mathtt{Vex}_q \max\{u, x\}(p^*, q^*) \geq \mathtt{Vex}_q \max\{u, x\}(p^*, q^*) \\ &\geq \sum_{r \in L} \mu^r \max\{u, x\}(p^*, q^*(r)) \geq \sum_{r \in L} \mu^r x(p^*, q^*(r)). \end{aligned}$$

However, by convexity in q:

$$y(p^*, q^*) \leq \sum_{r \in L} \mu^r y(p^*, q^*(r))$$

so that $(y - x)(p^*, q^*(r)) = \delta$ for all r, contradicting the extremality of (p^*, q^*).

∎

Corollary 4.17

$$\underline{w} = \overline{w} = x = y.$$

4.5.3 The functional equation on $\mathcal{C}(\Delta(K) \times \Delta(L))$

The purpose of this subsection is to extend the previous properties of the functional equations (a'), (b') to any continuous function u on $\Delta(K) \times \Delta(L)$.

Notations

Denote by $\mathcal{C}$ the set of continuous functions on $\Delta(K) \times \Delta(L)$ and by $\mathcal{U}$ the set of functions u arising from games: namely such that there exists a family $G^{k\ell}$ of $I \times J$ matrices with $u(p, q) = \mathtt{val}(\sum_{k \in K, \ell \in L} p^k q^\ell G^{k\ell})$.

Proposition 4.18

$\mathcal{U}$ is a vector space and a lattice.
$\mathcal{U}$ is dense in $\mathcal{C}$.

Proof
If u and u' belong to $\mathcal{U}$ then also:
i) $-u$ (take the opposite of the transpose of the matrices $G^{k\ell}$)
ii) λu, with $\lambda \geq 0$ (multiply by λ)
iii) $u + u'$ (take the product game with sets of moves $I \times I'$ and $J \times J'$ and payoff $B^{k\ell}_{ii'jj'} = G^{k\ell}_{ij} + G'^{k\ell}_{i'j'}$)
iv) $\max\{u, 0\}$ (add to the set I of moves of Player 1 a move with payoff identically 0.)
It is clear that $\mathcal{U}$ separates points (since it contains the affine functions) hence Stone-Weierstrass Theorem implies the density of $\mathcal{U}$ in $\mathcal{C}$. ∎

Definition
Given $u \in \mathcal{U}$, denote by $F(u)$ the function defined by Proposition 4.14, namely the only solution of

$$(a') \qquad f = \mathtt{Vex}_q \max\{u, f\}$$
$$(b') \qquad f = \mathtt{Cav}_p \min\{u, f\}.$$

Proposition 4.19
The mapping F from $\mathcal{U}$ to $\mathcal{C}$ has a unique continuous extension to $\mathcal{C}$. This extension is monotonic and non expansive.

Proof
Clearly F is monotonic and $F(u + c) = F(u) + c$ for any constant c. Hence F is nonexpansive on $\mathcal{U}$. Proposition 4.18 thus gives the result. ∎

Proposition 4.20
For any $u \in \mathcal{C}$, let

$$(a) \qquad f \geq \mathtt{Cav}_p\mathtt{Vex}_q \max\{u, f\}$$
$$(b) \qquad f \leq \mathtt{Vex}_q\mathtt{Cav}_p \min\{u, f\}.$$

$F(u)$ is the smallest solution of (a) and the largest solution of (b).
In particular $F(u)$ is the only solution of both.
Let also

$$(a') \qquad f = \mathtt{Vex}_q \max\{u, f\}$$
$$(b') \qquad f = \mathtt{Cav}_p \min\{u, f\}$$

$F(u)$ is the only solution of (a') and (b').

Proof
The operators $\mathtt{Cav}_p$, $\mathtt{Vex}_q$, max and min are continuous for the supremum norm hence $F(u)$ satisfies all properties since they are true on $\mathcal{U}$.
Uniqueness in (a'), (b') will follow from minimality in (a) as in Proposition 4.14. Let thus f satisfy (a) for some $u \in \mathcal{C}$. For $u_n \in \mathcal{U}$ with $u \geq u_n$, one obtains:

$$f \geq \mathtt{Cav}_p\mathtt{Vex}_q \max\{u, f\} \geq \mathtt{Cav}_p\mathtt{Vex}_q \max\{u_n, f\} = F(u_n).$$

So that by continuity $f \geq F(u)$. ∎

4.6 Recursive structure and operator approach

4.6.1 Recursive formula and dual games

This section is the counterpart of Subsection 3.5.1 in Chapter 3.

Proposition 4.21
v_n and v_λ satisfy the following recursive formula:

$$(n+1)v_{n+1}(p,q) = \mathbf{val}_{S^K \times T^L}\{\sum_{k\in K,\ell\in L} p^k q^\ell s^k G^{k\ell} t^\ell + n\sum_{i\in I,j\in J} \overline{s}(i)\overline{t}(j) v_n(p(i),q(j))\}$$

$$v_\lambda(p,q) = \mathbf{val}_{S^K \times T^L}\{\lambda\sum_{k\in K,\ell\in L} p^k q^\ell s^k G^{k\ell} t^\ell + (1-\lambda)\sum_{i\in I,j\in J} \overline{s}(i)\overline{t}(j) v_\lambda(p(i),q(j))\}$$

where $\overline{s} = \sum_{k\in K} p^k s^k$, $\overline{t} = \sum_{\ell\in L} q^\ell t^\ell$ and $\overline{p}(i)$ (resp. $\overline{q}(j)$) is the conditional probability distribution on K given i (resp. on L, given j).

Proof
The proof is the same in both cases.
We show that, for any strategy τ of Player 2, there exists a strategy σ of Player 1 satisfying:

$$(n+1)\gamma^{pq}_{n+1}(\sigma,\tau) \geq \min_{t\in T^L}\max_{s\in S^K} \rho_{n+1}(s,t)$$

where

$$\rho_{n+1}(s,t) = \{\sum_{k\ell} p^k q^\ell s^k G^{k\ell} t^\ell + n\sum_{ij} \overline{s}(i)\overline{t}(j) v_n(p(i),q(j))\}$$

In fact let τ_1 be induced by τ at stage 1. Player 1 computes the conditional probability on L after stage 1 and can then achieve, after history (i,j), the amount $v_n(p(i),q(j))$ in the remaining game. By maximizing on S^K the total payoff $\rho_{n+1}(s,\tau_1)$ we obtain the lower bound.
The result now follows from the existence of v_{n+1}.

∎

The compact writing of these equations is :

$$(n+1)v_{n+1}(p,q) = \mathbf{val}\{g_1(p,q) + nE[v_n(\widetilde{p},\widetilde{q})]\}$$

$$v_\lambda(p,q) = \mathbf{val}\{\lambda g_1(p,q) + (1-\lambda)E[v_\lambda(\widetilde{p},\widetilde{q})]\}$$

Note that contrary to the case of lack of information on one side, the recursive formula does not allow to construct optimal strategies, because none of the players knows the new **state variable** $(\widetilde{p},\widetilde{q})$.

However one can, like in the previous Chapter 3, introduce the dual game, but there are now two of them. Denote by $\mathcal{D}_1$ the operator of duality w.r.t. the first parameter k. The dual game $\mathcal{D}_1[G_n](x,q)$ of $G_n(p,q)$ (resp. $\mathcal{D}_1[G_\lambda](x,q)$ of $G_\lambda(p,q)$) is played as follows. Player 1 chooses k (Player 2 is not informed). ℓ is chosen according to q and Player 2 is informed (and not Player 1). Then at each stage m, knowing the previous history h_m, both players choose moves (i_m, j_m). The payoff is $\frac{1}{n}\sum_{m=1}^{n} G^{k\ell}_{i_m j_m} - x^k$ (resp. $\sum_{m=1}^{\infty}\lambda(1-\lambda)^{m-1}G^{k\ell}_{i_m j_m} - x^k$).
The value $w^1_n(x,q)$ of $\mathcal{D}_1[G_n](x,q)$ satisfies:

$$w^1_n(x,q) = \max_{p\in\Delta(K)}\{v_n(p,q) - \langle p,x\rangle\} = \Lambda_s(v_n(.,q))(x)$$

$$v_n(p,q) = \inf_{x\in\mathbb{R}^K}\{w^1_n(x,q) + \langle p,x\rangle\} = \Lambda_i(w^1_n(.,q))(p)$$

and similarly for $\mathcal{D}_1[G_\lambda](x,q)$:

$$w^1_\lambda(x,q) = \max_{p\in\Delta(K)}\{v_\lambda(p,q) - \langle p,x\rangle\} = \Lambda_s(v_\lambda(.,q))(x)$$

$$v_\lambda(p,q) = \inf_{x\in\mathbb{R}^K}\{w^1_\lambda(x,q) + \langle p,x\rangle\} = \Lambda_i(w^1_\lambda(.,q))(p).$$

In the other dual games $\mathcal{D}_2[G_n](p,y)$ and $\mathcal{D}_2[G_\lambda](p,y)$, the role of the players are exchanged and y is added to the payoff.

From the "primal" recursive formula (Proposition 4.21) one obtains:

Proposition 4.22

$$w^1_{n+1}(x,q) = \min_{t\in T^L}\max_{i\in I}\Big(\frac{n}{n+1}\Big)\sum\nolimits_{j\in J}\bar{t}(j)w_n\Big(\frac{n+1}{n}x - \frac{1}{n}\sum\nolimits_{\ell\in L}A^{\ell}_{ij}q^{\ell}(j), q(j)\Big)$$

$$w^1_\lambda(x,q) = \min_{t\in T^\ell}\max_{i\in I}(1-\lambda)\sum\nolimits_{j\in J}\bar{t}(j)w_\lambda\Big(\frac{1}{1-\lambda}x - \frac{\lambda}{1-\lambda}\sum\nolimits_{\ell\in L}G^{\ell}_{ij}q^{\ell}(j), q(j)\Big)$$

where G^ℓ_{ij} stands for the vector with components $\{G^{k\ell}_{ij}\}$, $k\in K$.

Proof
Using the recursive formula for v_{n+1} one has:

$$w^1_{n+1}(x,q) = \max_{p\in\Delta(K)}\max_{s\in S^K}\min_{t\in T^L}$$

$$\Big\{\frac{1}{n+1}[\sum\nolimits_{k\ell}p^k q^\ell s^k G^{k\ell} t^\ell + n\sum\nolimits_{ij}\bar{s}(i)\bar{t}(j)v_n(p(i),q(j))] - \langle p,x\rangle\Big\}.$$

Let $\pi = p\bullet\sigma$ in $\Delta(I\times K)$ (see Proposition 3.2) defined by $\pi(i,k) = p^k s^k(i)$. One has, π^i being the marginal on I, π_k the marginal on K and $\pi[i]$ the conditional on K given i:

$$w_{n+1}^1(x,q) = \max_{\pi\in\Delta(I\times K)} \min_{t\in T^L}$$

$$\left\{\frac{1}{n+1}[\sum_{k,i,\ell}\pi(i,k)G_i^{k\ell}t^\ell q^\ell + n\sum_{ij}\pi^i\bar{t}(j)v_n(\pi[i],q(j))] - \sum_k \pi_k x^k\right\}.$$

As in Proposition 3.20 one shows here that $\sum_i \pi^i v_n(\pi[i],q(j))$ is concave in π and $\sum_j \bar{t}(j)v_n(\pi[i],q(j))$ convex in t. Hence one obtains by the minmax theorem (e.g. Theorem A.7):

$$w_{n+1}^1(x,q) = \min_{t\in T^L} \max_{\pi\in\Delta(I\times K)}$$

$$\left\{\frac{1}{n+1}[\sum_{k,i,\ell}\pi(i,k)G_i^{k\ell}t^\ell q^\ell + n\sum_{ij}\pi^i\bar{t}(j)v_n(\pi[i],q(j))] - \sum_k \pi_k x^k\right\}.$$

Thus, using $t^\ell(j)q^\ell = \bar{t}(j)q^\ell(j)$ and the above formula expressing v_n in terms of w_n^1:

$$w_{n+1}^1(x,q) = \min_{t\in T^L} \max_{\pi\in\Delta(I\times K)} \left\{\frac{1}{n+1}[\sum_{k,i,\ell,j}\pi(i,k)G_{ij}^{k\ell}\bar{t}(j)q^\ell(j)\right.$$

$$\left. +n\sum_{ij}\pi^i\bar{t}(j) \inf_{\rho\in\mathbb{R}^K}(w_n(\rho,q(j)) + \langle\pi[i],\rho\rangle)] - \sum_k \pi_k x^k\right\}$$

$$= \min_{t\in T^L} \max_{\pi\in\Delta(I\times K)} \inf_{\rho\in\mathbb{R}^{K\times I\times J}} \left\{\frac{1}{n+1}[\sum_{k,i,\ell,j}\pi(i,k)G_{ij}^{k\ell}\bar{t}(j)q^\ell(j)\right.$$

$$\left. +n\sum_{ij}\pi^i\bar{t}(j)(w_n(\rho(i,j),q(j)) + \langle\pi[i],\rho(i,j)\rangle)] - \sum_k \pi_k x^k\right\}.$$

Again due to the convexity of w_n in ρ one can exchange the operators on ρ and π. Taking now extreme points for π one obtains:

$$w_{n+1}^1(x,q) = \min_{t\in T^L} \inf_{\rho\in\mathbb{R}^{K\times I\times J}} \max_{(i,k)\in(I\times K)}$$

$$\left\{\frac{1}{n+1}[\sum_{\ell,j} G_{ij}^{k\ell}\bar{t}(j)q^\ell(j) + n\sum_j \bar{t}(j)(w_n(\rho(i,j),q(j)) + \rho(i,j)^k] - x^k\right\}.$$

Finally:

$$w_{n+1}^1(x,q) = \min_{t\in T^L} \inf_{\rho\in\mathbb{R}^{K\times I\times J}} \max_{(i,k)\in(I\times K)}$$

$$\frac{n}{n+1}\sum_j \bar{t}(j)\left[w_n(\rho(i,j),q(j)) - (\frac{n+1}{n})x^k - \rho(i,j)^k - \frac{1}{n}\sum_\ell G_{ij}^{k\ell}q^\ell(j)\right]$$

and the result follows by choosing $\rho(i,j)$ equalizing in k.
The proof is similar for w_λ.

■

The above formula allows to define inductively the strategies trough the following state variables:
$q_1 = q$ and if $j_m = j$, $q_{m+1} = q_m(j)$;
$x_1 = x$ and given i_m and j_m, $x_{m+1} = \dfrac{n+1-m}{n-m}x - \dfrac{1}{n-m}\sum_\ell A^\ell_{i_m j_m} q^\ell_m(j)$.

Corollary 4.23
Player 2 has an optimal strategy in $\mathcal{D}_1[G_n](x,q)$ that depends only, at each stage m, on m, x_m and q_m (Markov on $\mathbb{R}^K \times \Delta(L)$).
Player 2 has an optimal strategy in $\mathcal{D}_1[G_\lambda](x,q)$ that depends only at each stage m on x_m and q_m (stationary on $\mathbb{R}^K \times \Delta(L)$).

Proof
Let t be optimal for $w^1_n(x,q)$ according to Proposition 4.22. Consider the strategy τ defined by: play t at stage 1 and then optimally in $\mathcal{D}_1[G_{n-1}](x_2,q_2)$. The proof is then similar to the one in Corollary 3.24.

∎

The relation between optimal strategies for Player 2 in primal and dual games allows to obtain the next result

Corollary 4.24
Player 2 has an optimal strategy in $G_n(p,q)$ that depends only, at each stage m, on m and x_m and q_m.
Player 2 has an optimal strategy in $G_\lambda(p,q)$ that depends only at each stage m on x_m and q_m.

Proof
Follows from Corollary 2.10 and Corollary 4.23.

∎

4.6.2 Operator approach

We use here the notations and results of Appendix C, Section 4 to obtain an alternative proof of existence and characterization for $\lim_{n\to\infty} v_n$ and $\lim_{\lambda\to 0} v_\lambda$.

Notation
$\mathcal{F}_s$ is the set of saddle functions on $\Delta(K)\times\Delta(L)$, separately continuous and uniformly bounded by $\|G\|$.
Using the notations of Appendix C.4 the operator $\mathbf{\Phi}(\alpha,f)$ corresponding to the value of the game $\Gamma(\alpha,f)$ is:

$$\Phi(\alpha, f)(p,q) = \mathtt{val}_{S^K \times T^L}\{\alpha\sum_{k\in K,\ell\in L} p^k q^\ell s^k G^{k\ell} t^\ell$$
$$+(1-\alpha)\sum_{i\in I,j\in J} \bar{s}(i)\bar{t}(j) f(p(i),q(j))\}$$

that will be shortly written as

$$\Phi(\alpha, f)(p,q) = \mathtt{val}\{\alpha g_1(p,q) + (1-\alpha)E(f(\tilde{p},\tilde{q}))\}$$

The incomplete information game is viewed here as a **stochastic game on the belief space** $\Delta(K)\times\Delta(L)$. Note that this embeding is based on the recursive formula. It thus relies on the minmax theorem and a priori does not extend to the infinite game.
The corresponding notations are $\omega = (p,q)\in\Delta(K)\times\Delta(L)$ for the state parameter, $X = \Delta(I)^K$ and $Y = \Delta(J)^L$ for the vectors of mixed moves, $g(x,y,\omega) = \sum_{k\ell} p^k q^\ell s^k G^{k\ell} y^\ell$ for the corresponding payoff and the transition $q(x,y,\omega)$ is defined for $\tilde{\omega} = (\tilde{p},\tilde{q}) = (p(i),q(j))$ by $q(x,y,\omega)(\tilde{\omega})$ equal to the probability $\bar{s}(i)\bar{t}(j)$.

Proposition 4.25
The operator $\Phi(\alpha,.)$ is well defined on $\mathcal{F}_s$ and maps $\mathcal{F}_s$ to itself.

Proof
Given f in $\mathcal{F}_s$

$$\Phi_{xy}(\alpha, f)(p,q) = \alpha\sum_{k\ell} p^k q^\ell s^k G^{k\ell} t^\ell + (1-\alpha)\sum_{ij} \bar{s}(i)\bar{t}(j) f(p(i),q(j))$$

is continuous in x. In addition the payoff appears as the expectation of the evaluation of a concave function at the posterior given i under the law x, hence is concave in x by Jensen's inequality. Thus Theorem A.7 applies to the game $\Gamma(\alpha, f)(p,q)$ and $\Phi(\alpha, f)(p,q)$ exists.
Consider now the function $(p,q)\mapsto\Phi(\alpha, f)(p,q)$. Given $\varepsilon > 0$, let η such that $\|p-p'\| \leq \eta$ implies $|f(p,q)-f(p',q)| \leq \varepsilon$. For $\delta > 0$ small enough, $\sum p^k s_i^k \geq \varepsilon$ and $\|p-p'\| \leq \delta$ imply $\|p(i)-p'(i)\| \leq \eta$. So that $\|p-p'\| \leq \min(\eta,\delta)$ implies:

$$|\Phi_{xy}(\alpha, f)(p,q) - \Phi_{xy}(\alpha, f)(p',q)| \leq \alpha\|G\|\min(\eta,\delta) + (1-\alpha)(2\|G\|\varepsilon + \varepsilon)$$

hence by non expansiveness:

$$|\Phi(\alpha, f)(p,q) - \Phi(\alpha, f)(p',q)| \leq \alpha\|G\|\min(\eta,\delta) + (1-\alpha)(2\|G\|\varepsilon + \varepsilon)$$

so that $\Phi(\alpha, f)(p,q)$ is separately continuous.
Finally the concavity in p of $\Phi(\alpha, f)(p,q)$ follows from the concavity in p of f and a splitting procedure. In fact let $p = \mu p_1 + (1-\mu)p_2$ be a convex combination. Given x_m optimal for Player 1 in $\Phi(\alpha, f)(p_m,q)$, $m = 1,2$,

consider x which chooses, if k, x_1 with probability $\frac{\mu p_1^k}{p^k}$. The corresponding payoff, given some y can be written as:

$$\begin{aligned}\mathbf{\Phi}_{xy}(\alpha,f)(p,q) &= \alpha(\mu\sum\nolimits_{k\ell}p_1^k q^\ell s_1^k G^{k\ell}t^\ell + (1-\mu)\sum\nolimits_{k\ell}p_2^k q^\ell s_2^k G^{k\ell}t^\ell)\\ &\quad +(1-\alpha)\sum\nolimits_{ij}\overline{s}(i)\overline{t}(j)f(\mu\frac{\overline{s}_1(i)}{\overline{s}(i)}p_1(i)+(1-\mu)\frac{\overline{s}_2(i)}{\overline{s}(i)}p_2(i),q(j))\\ &\geq \mu\mathbf{\Phi}_{x_1y}(\alpha,f)(p_1,q)+(1-\mu)\mathbf{\Phi}_{x_2y}(\alpha,f)(p_2,q).\end{aligned}$$

So that, for all y:

$$\mathbf{\Phi}_{xy}(\alpha,f)(p,q) \geq \mu\mathbf{\Phi}(\alpha,f)(p_1,q)+(1-\mu)\mathbf{\Phi}(\alpha,f)(p_2,q)$$

and the result follows. ∎

A first easy result is that any f in $\mathcal{F}_s$ is a fixed point of $\mathbf{\Phi}(0,.)$.

Lemma 4.26

$$\forall f\in\mathcal{F}_s,\ \mathbf{\Phi}(0,f)=f.$$

Proof
Recall that $\mathbf{\Phi}(0,f)(p,q)=\mathtt{val}_{X\times Y}E(f(\tilde{p},\tilde{q}))$ with $X=\Delta(I)^K$, $Y=\Delta(J)^L$ and that $(\tilde{p},\tilde{q})$ is a martingale with expectation (p,q).
For any $y\in\Delta(J)^L$ non-revealing, one has by Jensen's inequality:

$$\mathbf{\Phi}_{xy}(0,f)=E_x(f(\tilde{p},q))\leq f(p,q),\qquad \forall x\in X.$$

The dual inequality implies the result. ∎

In particular φ and φ^* (Propositions C.18 and C.20) coincide on $\mathcal{F}_s$.

Definition
Given f on $\Delta(K)$, let $\mathcal{E}_f$ denote the set of **extreme points** of f on $\Delta(K)$. Explicitly p belongs to $\mathcal{E}_f$ if the equality $f(p)=\sum_{r\in R}\lambda_r f(p_r)$ with $p=\sum_{r\in R}\lambda_r p_r$, R finite, $\lambda\in\Delta(R)$, $\lambda\gg 0$ and $p_r\in\Delta(K)$ implies $p=p_r$ for all r in R.
Similarly if f is defined on $\Delta(K)\times\Delta(L)$, $\mathcal{E}_f(q)$ denotes, for each $q\in\Delta(L)$, the set of extreme points of $f(.,q)$ on $\Delta(K)$.

Recall that $NR^1(p)$ denote the set of non-revealing strategies of Player 1. The proof of Lemma 4.26 also showed:

Corollary 4.27

$$NR^1(p)\subset X(0,p)$$
$$p\in\mathcal{E}_f(q)\Rightarrow NR^1(p)=X(0,f).$$

This means that to play non-revealing is optimal in the projective game $\Gamma(0, f)$ and is the only optimal strategy for Player 1 at each state (p, q) where f is strictly concave in p.
We now introduce two subsets of functions.

Definition
$\mathcal{A}^+$ consists of those f in $\mathcal{F}_s$, such that for any function h positive, concave and continuous on $\Delta(K)$, with $f + h$ strictly concave on $\Delta(K)$:

$$\varphi(f + h)(p, q) \leq 0, \qquad \forall (p, q) \in \Delta(K) \times \Delta(L).$$

In words, $\mathcal{A}^+$ corresponds to concave positive (in p) perturbations f^+ of f with $\varphi(f^+) \leq 0$.
A dual definition holds for $\mathcal{A}^-$ which is thus the set of functions f in $\mathcal{F}_s$, such that for any function h positive, concave and continuous on $\Delta(L)$ and with $f - h$ strictly convex on $\Delta(L)$:

$$\varphi(f - h)(p, q) \geq 0, \qquad \forall (p, q) \in \Delta(K) \times \Delta(L).$$

Proposition 4.28
$\mathcal{A}^+ \cap \mathcal{A}^-$ contains at most one continuous function.

Proof
Note simply that such a function belongs to the closure of $\mathcal{S}_0^+$ and apply Corollary C.26. ∎

We now turn to the study of the asymptotic behavior of the game and first deal with the discounted case. Let W be an accumulation point, as λ goes to 0, of the family $\{v_\lambda\}$, which is uniformly Lipschitz, hence relatively compact and let v_{λ_n} converge (uniformly) to W as $\lambda_n \to 0$. Note that W is Lipschitz.

Proposition 4.29

$$W \in \mathcal{A}^+.$$

Proof
Assume by contradiction: $\varphi(W + h)(p, q) \geq \delta > 0$ for some h positive, continuous and concave on $\Delta(K)$ with $W + h$ strictly concave. Use Corollary C.19 (or rather its dual) at (p, q) with $\rho = \delta/2$.
Thus given $y \in Y$:
a) Either there exists $x \in X$ with

$$\mathbf{\Phi}_{xy}(0, W + h)(p, q) \geq \mathbf{\Phi}(0, W + h)(p, q) + \eta.$$

Thus, a fortiori, since h is concave:

$$\boldsymbol{\Phi}_{xy}(0, W)(p,q) + h(p) \geq \boldsymbol{\Phi}(0, W+h)(p,q) + \eta \;=\; (W+h)(p,q) + \eta$$

by Lemma 4.26.
Hence by continuity, there exists N' such that for $n \geq N'$:

$$\boldsymbol{\Phi}_{xy}(\lambda_n, v_{\lambda_n})(p,q) \geq v_{\lambda_n}(p,q) + \eta/2.$$

b) Or there exists $x \in X(0, W+h)(p,q)$ with:

$$\varphi_{xy}(W+h)(p,q) \geq \varphi(W+h)(p,q) - \rho \geq \delta/2.$$

Note that $W + h$ being strictly concave, x is non revealing (Corollary 4.27), hence the stage payoff satisfies:

$$g(x, y; p, q) \geq E_{xy}((W+h)(\tilde{p}, \tilde{q})) + \delta/2 \geq W(p,q) + \delta/2$$

since W is convex in q. Thus there exists $N"$ such that for $n \geq N"$, since x is in $NR^I(p)$ and $v_{\lambda_n} \in \mathcal{F}_s$:

$$\boldsymbol{\Phi}_{xy}(\lambda_n, v_{\lambda_n})(p,q) \geq \lambda_n g(x, y; p, q) + (1-\lambda_n) v_{\lambda_n}(p,q) \geq v_{\lambda_n}(p,q) + \lambda_n \delta/4$$

Finally a) and b) imply that, for all $n \geq \max(N', N")$, and all y there exists x satisfying:

$$\boldsymbol{\Phi}_{xy}(\lambda_n, v_{\lambda_n})(p,q) > v_{\lambda_n}(p,q)$$

a contradiction to Property C.13. ∎
Obviously one has also $W \in \mathcal{A}^-$.

Consider now the finitely repeated game G_n. Recall the notations of section 4.1. $\underline{w} = \liminf_{n\to\infty} v_n$ and $\overline{w} = \limsup_{n\to\infty} v_n$. We first prove:

Proposition 4.30

$$\underline{w} = \overline{w}$$

Proof
See Exercise 8.7 a).

Denoting $V^* = \overline{w} = \underline{w}$ one obtains:

Proposition 4.31

$$V^* \in \mathcal{A}^+$$

Proof
See Exercise 8.7b).

The asymptotic properties follow:

Proposition 4.32
There exists a continuous function V in $\mathcal{A}^+ \cap \mathcal{A}^-$.
$\lim_{\lambda\to 0} v_\lambda$ *and* $\lim_{n\to\infty} v_n$ *exist and equal* V.

Proof
From Proposition 4.29, any accumulation point W of the family v_λ belongs to $\mathcal{A}^+ \cap \mathcal{A}^-$ and Proposition 4.28 shows that this set contains at most one continuous function. Hence v_λ converges to V. Similarly for v_n using Proposition 4.31. ∎

We recall the functional equations of Section 5:

$$(a') \qquad f = \mathtt{Vex}_q \max(u, f)$$

$$(b') \qquad f = \mathtt{Cav}_p \min(u, f)$$

The next result shows that on $\mathcal{F}_s$ the systems (a') and $\mathcal{A}^+$ (resp. (b') and $\mathcal{A}^-$) are equivalent.

Proposition 4.33
Let $f \in \mathcal{A}^-$ then f satisfies (a').

Proof
The proof is based on the following:

Lemma 4.34
Let $f \in \mathcal{A}^-$ and $p \in \mathcal{E}_f(q)$. Then:

$$f(p, q) \leq u(p, q).$$

Proof
Let h be strictly concave, continuous and positive on $\Delta(L)$, so that:

$$\varphi(f - h)(p, q) \geq 0.$$

Note that $X(0, f-h)(p,q) = X(0, f)(p,q) = NR^1(p)$ and $Y(0, f-h)(p,q) = NR^2(q)$ by Corollary 4.27. Hence:

$$\begin{aligned}\varphi(f-h)(p,q) &= \mathtt{val}_{NR^1(p) \times NR^2(q)}\{g(x, y; p, q) - E_{xy}((f-h)(\tilde{p}, \tilde{q}))\} \\ &= u(p,q) - (f-h)(p,q).\end{aligned}$$

So that $u(p,q) \geq f(p,q) - h(q)$ and the result follows. ∎

Another property is:

Lemma 4.35
Let g be a continuous concave function on $\Delta(K)$.

$$g(p) \leq u(p) \quad \forall p \in \mathcal{E}_g$$

is equivalent to:

$$g = \mathtt{Cav}_p \ \min(u, g).$$

Proof
In fact, g being concave in p, $g \geq \mathtt{Cav}_p \min(u, g)$.
On the other hand, g is smaller than $\min(u, g)$ at each point $p \in \mathcal{E}_g$, hence at each extreme point of its closed hypograph. It follows that $g \leq \mathtt{Cav}_p \min(u, g)$.
The other direction is easy: if $p \in \mathcal{E}_g$ the $\mathtt{Cav}_p$ operator is trivial hence $g(p) = \min(u, g)(p)$. ∎

Proposition 4.33 then follows from the two previous Lemmas. ∎

Note that Proposition 4.32 provides an alternative proof of existence of a solution to $\{(a')$ and $(b')\}$.
Finally we deduce the uniqueness of the solution of $\{(a')$ and $(b')\}$ on $\mathcal{F}_s$ trough the following:

Proposition 4.36
If $f \in \mathcal{F}_s$ satisfies (a') then it belongs to $\mathcal{A}^+$.

Proof
Let f be such a function and choose h positive, continuous concave on $\Delta(K)$ with $f+h$ strictly concave. Recall that $X(0, f+h)(p,q)$ is reduced to $NR^1(p)$.
On the other hand $Y(0, f+h)(p,q) = Y(0, f)(p,q)$. From (a') there exists a finite family q_r, $r \in R$, in $\Delta(L)$ and $\beta \in \Delta(R)$ such that:

$$f(p,q) = \sum_{r \in R} \beta_r \max(u, f)(p, q_r)$$

and

$$f(p, q_r) \geq u(p, q_r) \quad \forall r \in R.$$

Let y_r be a non-revealing optimal strategy for $u(p, q_r)$ and y^* the "splitting strategy" generating the q_r through the y_r's: if ℓ is the state, play y_r with probability $\beta_r q_r^\ell / q^\ell$.
Note that y^* is optimal for $\mathbf{\Phi}(0, f)(p,q)$ since, f being concave in p:

$$\sum_{j \in J} \overline{y}^*(j) f(p, q(j)) \geq \mathbf{\Phi}_{xy^*}(0, f)(p,q).$$

But f is convex in q and the $q(j)$'s can be decomposed as $q(j) = \sum_{r \in R} y_r(j) q_r$, hence:

$$f(p,q) \geq \sum_{r \in R} \beta_r f(p, q_r) \geq \sum_{j \in J} \overline{y}^*(j) f(p, q(j))$$

and the equality.
Let us now consider $\varphi(f+h)(p,q)$. For any x in $NR^1(p)$ one has:

$$\varphi_{xy^*}(f+h)(p,q) = g(x, y^*; p, q) - E_{x,y^*}(f+h)(\tilde{p}, \tilde{q})$$

But:

$$g(x, y^*; p, q) = \sum_{r \in R} \beta_r g(x, y_r; p, q_r)$$

and

$$E_{x,y^*}(f + h)(\tilde{p}, \tilde{q}) = \sum_{j \in J} \overline{y}^*(j) f(p, q(j)) + h(p).$$

Using $g(x, y_r; p, q_r) \le u(p, q_r)$ one obtains:

$$\varphi_{xy^*}(f + h)(p, q) \le \sum_{r \in R} \beta_r (u - f)(p, q_r) - h(p) \le -h(p)$$

and the result follows. ∎

We finally use Lemma 4.35 for a saddle function to obtain:

Proposition 4.37
Let f in $\mathcal{F}_s$. Then:

$$(a') \iff f(p, q) \ge u(p, q), \quad \forall q \in \mathcal{E}_f(p)$$

$$(b') \iff f(p, q) \le u(p, q), \quad \forall p \in \mathcal{E}_f(q).$$

4.7 Comments and extensions

4.7.1 Comments

a) As in the case of lack of information on one side the impact of the information on the payoff is through the **state variable** (p, q) (see Chapter 3, Subsection 3.7.1.a). This leads to introduce finite and discounted **splitting games** played as follows: at each stage m, knowing the state variable (p_m, q_m) player 1 (resp. 2) chooses a probability $\tilde{p}$ on $\Delta(K)$ with expectation p_m, $\tilde{p} \in \Delta^2_{p_m}(K)$ (resp. $\tilde{q}$ on $\Delta(L)$ with expectation q_m, $\tilde{q} \in \Delta^2_{q_m}(L)$), the payoff is the expectation of $u(\tilde{p}, \tilde{q})$ and the new state variable, chosen according to $(\tilde{p}, \tilde{q})$ (we identify the random variable and its law) is announced to the players. The recursive formula is then, for the value of the $n+1$ stage game:

$$(n+1) W_{n+1}(p, q) = \mathbf{val}_{\Delta^2_p(K) \times \Delta^2_q(L)} E\{u(\tilde{p}, \tilde{q}) + n W_n(\tilde{p}, \tilde{q})\}$$

and similarly in the discounted game:

$$W_\lambda(p, q) = \mathbf{val}_{\Delta^2_p(K) \times \Delta^2_q(L)} E\{\lambda u(\tilde{p}, \tilde{q}) + (1 - \lambda) W_\lambda(\tilde{p}, \tilde{q})\}$$

The corresponding compact game is played on $[0, 1]$, Player 1 (resp. 2) choosing a martingale $\{p_t\}$ with $p_0 = p$ (resp. $\{q_t\}$ with $q_0 = q$) and the payoff is:

$$E\left(\int_0^1 u(p_t, q_t) dt\right)$$

b) One of the main difference with the case of lack of information on one side lies the fact that the infinite game may have no value.
The proof in Section 3 shows that in this framework the best way to deal with private information looks as in the game "pick the largest integer": it is better to use his own information once the information of the other player is revealed. Hence a best reply is of the form: play non-revealing in order to exhaust the information of the opponent, then use one's own private information. Obviously, such strategies cannot in general generate an approximate saddle point.

c) The proof of Proposition 4.8 (existence of the limit of the values) is based on two aspects (see also Exercise 8.5):
- the existence of a solution f to (a') and (b')
- the evaluation of f given the state parameter (p_m, q_m).
Explicitely, an asymptotic good behavior for Player 1 is, if $u \geq f$ at (p_m, q_m),to play non-revealing optimally in $D(p_m, q_m)$, and otherwise to use first a splitting to reach points $\{p_m(r)\}$ where $u \geq f$ holds.
Note that this procedure requires the knowledge of q_m which is not available in the repeated game.
However an adapted procedure due to Heuer (1992a) that constructs a pseudo state variable $\overline{q}_m$ using approachability tools, allows to construct asymptotically optimal strategies, see Exercise 8.3.

d) The duality operators $\mathcal{D}_1$ and $\mathcal{D}_2$ introduced in section 6.1 replace the probability related to private information by a vector parameter. Starting with a game with lack of information on both sides $G(p,q)$, $\mathcal{D}_1[G](x,q)$ is thus a game with lack of information on one side where Player 2 is informed. One can then apply the duality operator $\mathcal{D}_2$ and this defines a game $\mathcal{D}_2 \circ \mathcal{D}_1[G](x,y)$. The operators actually commute and the "bidual game" $\mathcal{D}^2_{1,2}$ is well defined (Laraki, 2000c), see Exercise 8.6 .

e) A direct proof of the results of Section 5.3 not using game theoretical tools can be found in Mertens and Zamir (1977a).

f) One of the main differences between v_n and v_λ is the "elementary " structure of the former. In the current framework, v_n is piece wise bilinear, see Exercise 8.8 and compare Chapter 3, Exercise 8.2..

4.7.2 Signals

As in Chapter 3, Subsection 3.7.3 one can extend the model and consider the game with a signalling structure.
Level 2 corresponds to $I \times J$ information matrices for each player with values in some finite alphabet (A or B) or more generally in $\Delta(A \times B)$. One then defines non-revealing strategies sets $NR^1(p)$ and $NR^2(q)$ for each player and the non-revealing game $D(p,q)$ on the product set with value $u(p,q)$. The general analysis was introduced by Mertens (1972) and most of the results of this Chapter extend, especially Theorem 4.6, Mertens and Zamir (1977) and Propositions 4.7 and 4.10, Mertens (1972). One of the main ideas is to

introduce a majorant (and dually a minorant) game where Player 1 knows the posterior probability that Player 2 can compute and then to show that their asymptotic values coincide. (Note that the recursive formula holds in the majorant game in term of max min).
There is a cost to get information but we obtain the same speed of convergence as in the case with lack of information on one side: $\|v_n - \lim v_n\| \leq \frac{1}{\sqrt[3]{n}}$ (MSZ, Chapter VI) and recall that it is the best bound (Chapter 3, 7.4).
In the framework of level 3, namely type dependent signalling matrices, only partial results are yet available, Sorin (1985b). Among the open questions are: existence and characterization of *minmax*, *maxmin*, $\lim_{n\to\infty} v_n$ and $\lim_{\lambda\to 0} v_\lambda$.
The relation with stochastic games is discussed in Chapter 6, Section 6.5.8

4.7.3 Dependent case

This corresponds to the case where the private informations of the players are not independent.
The model and the results described below are due to Mertens and Zamir (1971-72) and (1980).
K is a finite set, G^k, $k \in K$ is a family of $I \times J$ matrices and $p \in \Delta(K)$ is an initial probability. In addition two partitions $K^1 = \{K_a^1\}$ and $K^2 = \{K_b^2\}$ of K are given, representing the information structure of the players on the state.
k is chosen according to p then a (resp. b) is announced to Player 1 (resp. 2) with $p \in K_a^1 \cap K_b^2$. Then the game is played like in Section 1.
The previous approach, called the independent case, corresponds to the model where K is a product $L_1 \times L_2$, p is a product probability $p_1 \otimes p_2$ with $p_i \in \Delta(L_i)$ and K^1 (resp. K^2) is the partition induced by the first (resp. second) component.
The main difference with the previous analysis is the range of the posterior distributions. A function on K is 1-measurable if it is constant on each component K_a^1 (and similarly for 2-measurable). Define:

$$\Pi_1(p) = \{q \in \Delta(K); q^k = \alpha^k p^k, \forall k \in K, \text{ and } \alpha \text{ is 1-measurable}\}$$

$$\Pi_2(p) = \{q \in \Delta(K); q^k = \alpha^k p^k, \forall k \in K, \text{and } \alpha \text{ is 2-measurable}\}$$

A real function f on $\Delta(K)$ is called 1-concave if for each $p \in \Delta(K)$, its restriction to $\Pi_1(p)$ is concave and similarly for 2-convex.
Finally for each real function f on $\Delta(K)$, $\mathtt{Cav}_1 f$ (resp. $\mathtt{Vex}_2 f$) denotes the smallest 1-concave function above f on $\Delta(K)$ (resp. the largest 2-convex function below f on $\Delta(K)$).
The splitting property extends to martingales in $\Pi_1(p)$ which is the set of posterior distributions that Player 1 can generate. The structure of the proofs is then roughly similar but the analysis is much more involved because the

process $\{p_m\}$ is controlled by both players. The results for the asymptotic approach are:

$$\lim_{n\to\infty} v_n = \lim_{\lambda\to 0} v_\lambda = v$$

exists and is the unique solution of:

$$v = \mathtt{Cav}_1 \min\{u, v\}$$

$$v = \mathtt{Vex}_2 \max\{u, v\}.$$

Concerning the infinite game one has the analog of Theorem 4.6:

$$minmax = \overline{V} = \mathtt{Vex}_2\mathtt{Cav}_1 u$$

$$maxmin = \underline{V} = \mathtt{Cav}_1\mathtt{Vex}_2 u.$$

The direct analysis of the functional equations, see Section 5.3, can also be extended, Sorin (1984b).
Finally the piecewise bilinearity of v_n (Section 7.1 f)) extend to the dependent case (Ponssard and Sorin, 1980b).

4.7.4 Sequential games

Assume that each matrix $G^{k\ell}$ corresponds to the payoffs of a sequential game where Player 1 plays first, i.e. chooses $i \in I$, then Player 2 chooses $j \in J$, knowing i. The value of the non-revealing game is now $u(p,q) = \max_{i\in I} \min_{j\in J} \sum_{k,\ell} p^k q^\ell G^{k\ell}_{ij}$.
The specific structure of the game allows to obtain more precise results:
- the recursive formula can be written as (Ponssard, 1975):

$$v_{n+1}(p,q) = \mathtt{Cav}_p \max_{i\in I} \mathtt{Vex}_q \min_{j\in J} \{\sum\nolimits_{k\in K, \ell\in L} p^k q^\ell G^{k\ell}_{ij} + n v_n(p,q)\}$$

- in the case of lack of information on one side, the sequence v_n is constant (Ponssard and Zamir, 1973)
- in the case of lack of information on both sides, the sequence v_n is increasing (Sorin, 1979)
from which one deduces that the speed of convergence is bounded by $O(1/n)$ and this is the best bound (Sorin, 1979).
In addition a sequential procedure to compute optimal strategies in the form of a pair of probability and a vector payoff is available (Ponssard and Sorin, 1982).

4.7.5 Incomplete information on one and half sides

Sorin and Zamir (1985) introduced the following game: there are two matrices A and B and $G(p_r)$, $r = 1, 2$, corresponds to the usual game with incomplete information on one side where p_r is the probability of A and Player 1 is informed. However $G(p_r)$, $r = 1, 2$ is itself chosen at random according to some distribution ρ. ρ is known by both players but only Player 2 knows r. To summarize, Player 1 will know the state (A or B) but not the beliefs of his opponent on it (p_r). This situation is called **lack of information on one and half sides**. Formally it reduces to a game with lack of information on both sides and the following structure where Player 1 is told the row and Player 2 the column:

A	A
B	B

Payoffs

$p_1\rho_1$	$p_2\rho_2$
$(1-p_1)\rho_1$	$(1-p_2)\rho_2$

Probabilities

Note that the probability distribution is not the product of its marginals, hence corresponds to the dependent case.
In such games the maxmin may differ from the minmax, meaning that they really behave differently from games with lack of information on one side.
An illuminating presentation is in Aumann and Maschler (1995), pp. 145-154.

4.7.6 Transcendental values

An important distinction between the asymptotic and the uniform approach is related to the "algebraic properties" of the value. Recall that $u(p,q)$ is the value of a finite game with entries that are bilinear functions of (p,q), hence u is a rational fraction of (p,q). In particular if all games $G^{k\ell}$ have rational (or algebraic) coefficients $u(p,q)$ will be algebraic at rational points. This property is preserved by the $\mathtt{Cav}$ and $\mathtt{Vex}$ operators so that, by Theorem 4.6 the min max and the maxmin are also algebraic. However Mertens and Zamir (1981) have produced an example where the solution of $\{(a) \text{ and } (b)\}$ (Proposition 4.20), hence $v(p,q) = \lim_{n\to\infty} v_n(p,q) = \lim_{\lambda\to 0} v_\lambda(p,q)$ has transcendental value for some rational entries.
This also shows an important difference with stochastic games (Chapter 5) where $\lim_{\lambda\to 0} v_\lambda$ is "algebraic".
When mixing both aspects (stochastic games with incomplete information, Chapter 6, Section 6.5) one even obtains games with "transcendental" maxmin or minmax.

4.8 Exercises

4.8.1 Complement to 4.8

Prove the assertion in Proposition 4.8.
(If $f < \mathtt{Vex}_q\mathtt{Cav}_p \min\{u, f\}$ at some point (p,q) one can strictly increase f and it will still satisfy the large inequality).

4.8.2 CavVex is saddle

Prove that for any real function f on $\Delta(K)\times\Delta(L)$, $\mathtt{Cav}_p\mathtt{Vex}_q f$ is saddle. (Use that for any function g on $\Delta(K)$, $\mathtt{Cav}_p g(p) = \sup\{\sum_{r\in R}\lambda_r g(p_r); \lambda\in\Delta(R), p_r\in\Delta(K), p=\sum_{r\in R}\lambda_r p_r, R \text{ finite}\}$.)

4.8.3 Approachability in asymptotic games, Heuer (1992a)

a) Let v be a solution of (a') and (b') (Proposition 4.14).
Let $B(q) = \{\beta\in\mathbb{R}^K; \langle\beta,p\rangle \geq v(p,q), \forall p\in\Delta(K)\}$ and $B(p,q) = \{\beta\in B(q); \langle\beta,p\rangle = v(p,q)\}$ be the set of supporting hyperplane to $v(.,q)$ at p.
Show that if $q(r)$ is a splitting for v at (p,q) (i.e. there exist $\lambda\in\Delta(R)$, $q(r)\in\Delta(L)$, $q=\sum_r\lambda_r q(r)$, $v(p,q)=\sum_r\lambda_r v(p,q(r))$ and β belongs to $B(p,q)$ there is a representation $\beta=\sum_r\lambda_r\beta(r)$ with $\beta(r)\in B(q(r))$.
b) We now consider the game $G_n(p,q)$ and will define inductively a strategy of Player 2. Let $\zeta(h_m,i_m)=\zeta_m$ be the conditional vector payoff from the point of view of Player 2 at stage m if the move of Player 1 was i_m. Explicitely

$$\zeta^k(h_m,i_m) = \sum_\ell q^\ell(h_m) G^{k\ell}_{i_m j}\tau^\ell(h_m)[j].$$

Show that $n\overline{\gamma}_n(\sigma,\tau) = E_\sigma(\langle p_n, \sum_{m=1}^n \zeta_m\rangle)$.
c) Like in approachability theory (Appendix B) Player 2 will control the vector payoff and consider the expectation of $\frac{1}{n}\sum_{m=1}^n\zeta_m$ with respect to any realization of a strategy of Player 1. We now describe τ.
At stage one choose $\beta=\beta_1\in B(p,q)$ and play optimally in $D(p,q)$. Define now $q_2=q_1=q$ and ξ_2 by

$$n\beta = \zeta_1 + (n-1)\xi_2.$$

In words, β is the original objective and ξ_2 is the new one (normalized) after stage one for the remaining $n-1$ stage game. Assume now ξ_m and q_m defined before stage m.
- if $\xi_m\notin B(q_m)$, let β_m be the projection of ξ_m on the convex set $B(q_m)$. π_m, the vector in $\Delta(K)$ parallel to $\beta_m-\xi_m$ will play the rôle of the unknown posterior. Note that $\beta_m\in B(\pi_m,q_m)$. There are now two subcases:
(i) If $u(\pi_m,q_m)\leq v(\pi_m,q_m)$, Player 2 plays optimally in $D(\pi_m,q_m)$. The new parameters are $q_{m+1}=q_m$ and ξ_{m+1} with $(n-m+1)\xi_m = \zeta_m+(n-m)\xi_{m+1}$.
(ii) Otherwise, let $q_m(r)$ be a splitting for v at (π_m,q_m) with in addition $u(\pi_m,q_m(r))\leq v(\pi_m,q_m(r))$ and $\beta_m(r)$ associated to β_m. Play now, given $q_m(r)$, optimally in $D(\pi_m,q_m(r))$. The new parameters are $q_{m+1}=q_m(r)$ and ξ_{m+1} with

$$(n-m+1)(\xi_m+\beta_m(r)-\beta_m) = \zeta_m + (n-m)\xi_{m+1}.$$

Somehow after the splitting the objective is redefined before the actual play.
- if $\xi_m\in B(q_m)$, τ is an arbitrary NR strategy at this stage and we follow case

(i) above.
Let $\mathcal{H}_m^+$ denotes the filtration before stage m but after the splitting. Show that:

$$E(\langle\pi_m, \zeta_m\rangle|\mathcal{H}_m^+) = E(\sum_{k\ell} \pi_m^k q_m^\ell(r) G_{i_m j_m}^{k\ell}) \leq u(\pi_m, q_m(r))$$

$$\leq v(\pi_m, q_m(r)) \leq \langle\pi_m, \beta_m(r)\rangle$$

hence

$$E(\langle\pi_m, \zeta_m\rangle|\mathcal{H}_m) = \sum_r \lambda_r E(\langle\pi_m, \zeta_m\rangle|\mathcal{H}_m^+) \leq \sum_r \lambda_r \langle\pi_m, \beta_m(r)\rangle \leq \langle\pi_m, \beta_m\rangle.$$

In particular one has:

$$E(\langle\beta_m - \xi_m, \zeta_m - \beta_m\rangle|\mathcal{H}_m) \leq 0. \qquad (**)$$

Note that $E(\xi_m + \beta_m(r) - \beta_m)|\mathcal{H}_m) = E(\xi_m|\mathcal{H}_m)$ so that:

$$n\beta = E(\zeta_1 + \ldots + \zeta_m + (n-m)\xi_{m+1}).$$

Hence in order to control $\frac{1}{n}\sum_{m=1}^n \zeta_m$ it is enough to control $\frac{1}{n}\xi_{m+1}$. To do this we evaluate inductively the distance from ζ_m to the set $B(q_m)$. Explicitely:

$$E(E[\|\xi_{m+1} - \beta_{m+1}\|^2|\mathcal{H}_m^+]|\mathcal{H}_m) \leq E(E\|\xi_{m+1} - \beta_m(r)\|^2|\mathcal{H}_m^+]|\mathcal{H}_m)$$

$$\leq E(E\|\frac{1}{n-m}((n-m+1)(\xi_m - \beta_m) - (\zeta_m - \beta_m(r))\|^2|\mathcal{H}_m^+]|\mathcal{H}_m)$$

$$\leq \frac{(n-m)^2}{(n-m+1)^2}\|\xi_m - \beta_m\|^2$$

$$-\frac{2}{n-m}E(E[\langle\xi_m - \beta_m, \zeta_m - \beta_m(r)\rangle|\mathcal{H}_m^+]|\mathcal{H}_m)$$

$$+E(E\|\zeta_m - \beta_m(r))\|^2|\mathcal{H}_m^+]|\mathcal{H}_m).$$

Deduce by induction, using (**), that $d_m^2 = E(\|\xi_m - \beta_m\|^2)$ satisfies:

$$d_m^2 \leq \frac{4m\|G\|^2}{n+1-m}.$$

Conclude that there exists a constant M with $E(\|\xi_n\|) \leq M\sqrt{n}$, hence, for all σ:

$$E(\frac{1}{n}\sum_{m=1}^n \zeta_m) \leq \beta + \frac{M}{\sqrt{n}}$$

so that:

$$v_n(p,q) \leq \langle p, \beta\rangle + \frac{M}{\sqrt{n}} = v(p,q) + \frac{M}{\sqrt{n}}.$$

d) Extend the result to the generalized game:

$$\beta = E(\sum_{m=1}^{r} \mu(m)\zeta_m + (1 - \sum_{m=1}^{r} \mu(m))\xi_{r+1}$$

(with speed of convergence of $\mu(1)^{1/2}$).
e) Extend the result to the dependent case.

4.8.4 Alternative proof for $\lim v_\lambda$, Laraki (2001)

a) Deduce from Proposition 4.21 that:

$$v_\lambda(p,q) \leq \max_{s \in X_\lambda(p,q)} \min_{t \in T} \{\lambda \sum_{k\ell} p^k q^\ell s^k G^{k\ell} t + (1-\lambda) \sum_i \bar{s}(i) v_\lambda(p(i), q)\}$$

where $X_\lambda(p,q)$ is the set of optimal strategies of Player 1 in $G_\lambda(p,q)$. Hence by concavity in p:

$$\max_{s \in X_\lambda(p,q)} \min_{t \in T} \{\sum_{k\ell} p^k q^\ell s^k G^{k\ell} t - \sum_i \bar{s}(i) v_\lambda(p(i), q)\} \geq 0$$

b) Let w be an accumulation point of the family v_λ as λ goes to 0 and let X^* be the corresponding set of accumulation points of $X_\lambda(p,q)$ in X. Note that $w = \mathtt{val} E[w(\tilde{p}, \tilde{q})]$ and that $X^* \subset X_0(w)$, set of optimal strategies of Player 1 in the game with payoff $E[w(\tilde{p}, \tilde{q})]$. Hence:

$$\max_{X_0(w)} \min_{T} \{\sum_{k\ell} p^k q^\ell s^k G^{k\ell} t - \sum_i \bar{s}(i) w(p(i), q)\} \geq 0$$

Recall (Corollary 4.27) that if $p \in \mathcal{E}_w(q)$ one has $X_0(w) \subset NR^1(p,q)$ so that:

$$u(p,q) - w(p,q) \geq 0.$$

c) Assume that w_1 (resp. w_2) is a saddle continuous function on $\Delta(K) \times \Delta(L)$ such that $p \in \mathcal{E}_w(q)$ implies $u(p,q) \geq w_1(p,q)$ (and a dual property for w_2). Show that $w_2 \geq w_1$.
(Consider an extreme point of the set where $w_1 - w_2$ is maximal, compare Proposition 4.16).
d) Deduce that $\lim v_\lambda$ exists.

4.8.5 Generalized game

a) Let v satisfy (a') and (b') (Proposition 4. 14). Deduce from the proof of Proposition 4.8 and from Proposition 4.14 that in $\mathcal{G}(p,q)$, for any τ, Player 1 has a strategy σ such that:

$$\rho_m(\sigma,\tau)(h_m) \geq X_m - \|G\| LV(\tau, q)$$

where X_m is a $\mathcal{H}_m$ submartingale with $X_1 = v$.
b) Show that in a game with generalized payoff $\gamma_\mu(\sigma,\tau) = \sum_m \mu(m)\gamma_m(\sigma,\tau)$ one has:

$$v_\mu(\sigma,\tau) \geq v - \|G\| L^{1/2} \mu(1)^{1/2}.$$

4.8.6 The bidual game, Laraki (2000c)

We consider games with lack of information on both sides in an abstract set up like in Chapter 2. A family $G^{k\ell}$ of continuous bilinear maps from $S \times T$ to $\mathbb{R}$ are given. To each $(p,q) \in \Delta(K) \times \Delta(L)$ is associated a game with lack of information on both sides $G(p,q)$ as usual.
a) Let $\mathcal{D}_1$ (resp. $\mathcal{D}_2$) denote the duality operator with respect to k (resp. ℓ). Show that both games $\mathcal{D}_1 \circ \mathcal{D}_2[G](x,y)$ and $\mathcal{D}_2 \circ \mathcal{D}_1[G](x,y)$ have the same max min and min max.
b) Define for any $z \in \mathbb{R}^{K \times L}$ the bidual game $\mathcal{D}^2[G](z)$ as follows:
Player 1 chooses k and Player 2 chooses ℓ.
Both players play σ and τ like in G.
The payoff is $E_{\sigma,\tau}(G^{k\ell}(s,t) - z^{k\ell})$.
Given $x \in \mathbb{R}^K$ and $y \in \mathbb{R}^L$ let $z = x \oplus y$ be defined by $z^{k\ell} = x^k + y^\ell$. Show, for example for the minmax $\overline{v}$ of G and $\overline{w}$ of $\mathcal{D}^2[G]$, that:

$$\overline{w}(z) = \mathtt{val}_{\Delta(K) \times \Delta(L)}\{\overline{v}(p,q) - \sum_{k\ell} p^k z^{k\ell} q^\ell\}$$

$$\overline{v}(p,q) = \mathtt{val}_{\mathbb{R}^K \times \mathbb{R}^L}\{\overline{w}(x \oplus y) + \sum_k p^k x^k + \sum_\ell q^\ell y^\ell\}$$

and that $\overline{w}$ is determined by the values $\overline{w}(x \oplus y)$.

4.8.7 Propositions 4.30 and 4.31

a) By contradiction assume $\overline{w} \neq \underline{w}$. Deduce that there exist $\overline{h}$ and $\underline{h}$ both continuous and strictly concave on $\Delta(L)$ (resp. $\Delta(K)$), a point (p,q) and $\delta > 0$ with:

$$\varphi(\mathtt{Vex}_q(\underline{w} + \underline{h}))(p,q) - \varphi(\mathtt{Cav}_p(\overline{w} - \overline{h}))(p,q) \geq \delta$$

and $\mathtt{Vex}_q(\underline{w} + \underline{h}))(p,q) = (\underline{w} + \underline{h})(p,q)$, $\mathtt{Cav}_p(\underline{w} + \underline{h}))(p,q) = (\underline{w} + \underline{h})(p,q)$. One can assume $\varphi(\mathtt{Vex}_q(\underline{w} + \underline{h}))(p,q) \geq \delta/2$. Proceed then as in Proposition 4.29 by using Corollary C.19 to prove the existence of positive constants α and γ such that $|v_{n+1}(p,q) - \underline{w}(p,q)| \leq \gamma$ implies:

$$(n+1)v_{n+1}(p,q) \geq (n+1)v_n(p,q) + \alpha$$

Deduce that the sequence $v_n(p,q)$ cannot have $\underline{w}(p,q)$ as accumulation point by considering a large index m where v_m belongs again to the intervall $|v_m(p,q) - \underline{w}(p,q)| \leq \gamma/2$.
b) From a) $w = \underline{w} = \overline{w}$ is saddle. Assume $\varphi(w + h)(p,q) > \delta$ with h positive and strictly concave on $\Delta(K)$ and $\delta > 0$. Obtain a contradiction by proceeding as in a).

4.8.8 $v_n(p,q)$ piecewise bilinear, Ponssard and Sorin (1980a)

Consider the repeated game $G_n(p,q)$. Let A (resp. B) be the finite set of pure strategies of Player 1 (resp. 2) restricted to $H_m, m \leq n$ and $C^{k\ell}_{ab}$ the payoff corresponding to the couple (a,b) in state (k,ℓ). Show that $v_n(p,q)$ is the value of the following dual linear programs

$$\begin{array}{ll} \max \sum_\ell q^\ell \alpha^\ell & \\ \sum_{a,k} \pi^k_a C^{k\ell}_{ab} \geq \alpha^\ell \ \forall b, \forall \ell & \\ \sum_a \pi^k_a = p^k & \forall k \\ \pi^k_a \geq 0 & \forall a, \forall k \end{array} \qquad \begin{array}{ll} \min \sum_\ell p^k \beta^k & \\ \sum_{b,\ell} \rho^\ell_b C^{k\ell}_{ab} \leq \beta^k \ \forall a, \forall k & \\ \sum_b \rho^\ell_b = q^\ell & \forall \ell \\ \rho^\ell_b \geq 0 & \forall b, \forall \ell \end{array}$$

Deduce that v_n is concave convex.
Prove that v_n is piecewise bilinear in the following sense: there exist finite partitions of $\Delta(K)$ (resp. $\Delta(L)$) into convex polyhedras P_i (resp Q_j) such that the restriction of v_n to each product $P_i \times Q_j$ is bilinear.
For the extension to the dependent case, Section 7.3, see Ponssard and Sorin (1980b).

4.8.9 Sequential games

We consider the framework of Section 7.4.
a) Prove that the recursive formula can be written as :

$$v_{n+1}(p,q) = \mathtt{Cav}_p \max_i \mathtt{Vex}_q \min_j \{\sum\nolimits_{k,\ell} p^k q^\ell G^{k\ell}_{ij} + n v_n(p,q)\}.$$

b) Use the properties $\mathtt{Vex}(f+g) \geq \mathtt{Vex}\, f + \mathtt{Vex}\, g$ and $\mathtt{Cav}(f + \alpha \mathtt{Cav}\, f) = (1+\alpha)\, \mathtt{Cav}\, f$ $(\alpha > 0)$ to show that the sequence v_n is increasing.
Deduce from Proposition 3.22 (Chapter 3) that for sequential games with incomplete information on one side where the informed player plays first the sequence v_n is constant.
c) Let $f(p,q) = -(\sum_i \min_j \sum_{k\ell} p^k q^\ell G^{k\ell}_{ij}) + C$ where C is a constant such that $v_1 \geq v + f$ (where v satisfies proposition 4.14). Show by induction $n v_n \geq n v + f$ and deduce a speed of convergence of the order of $1/n$.

4.9 Notes

The model of this chapter and the first results were provided by Aumann, Maschler and Stearns in 1967, see Aumann and Maschler (1995), Chapters 2 and 3.
The informational content of a strategy and the results of Section 2 are due to Stearns. Section 3 follows Aumann and Maschler.
The analysis of the compact case (Section 4) is due to Mertens and Zamir (1971-72) as well as Section 5 (see also Mertens and Zamir (1977a)).
We followed Rosenberg (1998) in Section 6.1 and Rosenberg and Sorin (2001) in 6.2.

5 Stochastic games

5.1 Presentation

A stochastic game is a repeated game where the state changes from stage to stage according to a transition depending on the current state and the moves of both players.
The game is thus specified by a **state space** Ω, move sets I and J (that could depend upon Ω), a real bounded payoff function g defined on $I \times J \times \Omega$ and a transition probability q from $I \times J \times \Omega$ to Ω. All sets under consideration are finite.
The standard hypothesis that will be mainly considered here assumes that the play, including the current stage, is public, hence the game form is as follows. An initial state ω_1 is given and known by the players. Inductively at stage n, knowing the past **history** h_n of moves and states, $h_n = (w_1, i_1, j_1, \omega_2, ..., i_{n-1}, j_{n-1}, w_n)$, both players choose a move, $i_n \in I$ for Player 1, $j_n \in J$ for Player 2 and a new state w_{n+1} is selected according to the distribution $q(.|i_n, j_n, \omega_n)$ on Ω. The payoff at stage n is $g_n = g(i_n, j_n, \omega_n)$. A play induces a stream of payoffs $\{g_n\}_{n=1}^{\infty}$.
Let $H_n = (\Omega \times I \times J)^{n-1} \times \Omega$ be the set of histories at stage n and $H = \cup_{n \geq 1} H_n$ the set of all histories. As in the previous chapters one considers behavioral strategies, i.e. mappings from H to probabilities on I (resp. on J), for Player 1 (resp. Player 2). (See Appendix D). A strategy is **Markov** if it depends only on the stage and on the current state and **stationary** if it depends only on the current state (in this last case it is defined by a mapping from states to mixed moves). As in other repeated games (see Chapter 3, Section 3.1) there are several payoffs associated to the stochastic game form . We denote by $G_n(\omega)$, resp. $G_\lambda(\omega)$ and $G_\infty(\omega)$ the finite, discounted and infinite games starting from state ω and by $v_n(\omega)$, $v_\lambda(\omega)$, $v_\infty(\omega)$ the corresponding values.

Contrary to the case of incomplete information where the finiteness of the parameter space was crucial, several results hold here for an uncountable state space. In this case Ω is equipped with a σ−field $\mathcal{C}$. The case of compact move spaces will also be considered. In both extensions, H inherits a natural measurable product structure. However, for the payoff to be well defined we shall consider only measurable strategies (but other choices are possible: universally measurable,....). Section 2 is devoted to the basic recursive structure of stochastic games: existence of a value and properties of optimal strategies in the compact case (G_n and G_λ) follow.

In Section 3 a famous game ("Big Match") is studied and several tools are described for this specific example. The next Section 4 exhibits the algebraic structure available in the **finite setup** where move and state spaces are finite, and presents the consequences for the asymptotic properties of v_n and v_λ. Section 5 still considers asymptotic properties but for absorbing games (where the state can change at most once) with compact move sets, using the operator approach (Appendix C.4). The results of Section 4 are used in Section 6 to study the infinite game G_∞ and to prove the main result: existence of a uniform value in the finite case. Recall finally that when there are no players, a stochastic game reduces to a Markov chain, and when only one player is present ($\#J = 1$) the framework is of Markov decision processes (or dynamic programming). Section 7 describes additional results available in these cases: some of them do not extend to stochastic games, for some others the problem is open. All the previous analysis is done in the standard case where the triple (i_n, j_n, ω_n) is announced to both players after stage n. For the more general class with signals, see Exercises 9.5 and 9.6 and Chapter 6, Section 6.4.

5.2 Compact games: existence

Since the state is known by both players the **recursive approach** is basic in stochastic games (see Appendix C, Section 4A). It relies on the **auxiliary one shot game** with state dependent terminal payoff and the associated **Shapley operator**.

5.2.1 The auxiliary game

First few notations are needed. $X = \Delta(I)$, $Y = \Delta(J)$ denote the sets of mixed moves. To each f in the set $\mathcal{F}$ of real bounded measurable functions on $(\Omega, \mathcal{C})$ one associates an auxiliary one shot game $\mathbf{G}(f)$ with strategy sets X and Y and payoff in state ω given by:

$$\Psi_{xy}(f)(\omega) = g(x, y, \omega) + E_{q(x,y,\omega)}(f)$$

where

$$g(x, y, \omega) = \int_{I \times J} g(i, j, \omega) x(di) y(dj)$$

is the current expected payoff at ω under x and y and

$$E_{q(x,y,\omega)}(f) = \int_{I \times J \times \Omega} f(\omega') q(d\omega' | i, j, w) x(di) y(dj)$$

is the expectation of the future payoff defined by the function f on the state space. In the finite setup, (I, J and Ω finite) the existence of a value follows from the minmax Theorem A.5. This defines the **Shapley operator** $\mathbf{\Psi}$: $f \mapsto \mathbf{\Psi}(f)$ through the relation:

$$\mathbf{\Psi}(f)(\omega) = \mathtt{val}\mathbf{\Psi}_{xy}(f)(\omega).$$

One can also consider for each α in $[0,1]$ the operator $\mathbf{\Phi}(\alpha, .)$ on $\mathcal{F}$ defined by:

$$\mathbf{\Phi}(\alpha, f) = \alpha \mathbf{\Psi}(\frac{(1-\alpha)}{\alpha} f).$$

It corresponds to the value of the game $\Gamma(\alpha, f)$ on $X \times Y$ with payoff

$$\mathbf{\Phi}_{xy}(\alpha, f)(\omega) = \alpha g(x, y, \omega) + (1-\alpha) E_{q(x,y,\omega)}(f).$$

Again, in the finite setup the existence of v_n and v_λ are immediate and one has:

$$v_\lambda(\omega) = \mathtt{val}_{X \times Y} \mathbf{\Phi}_{xy}(\lambda, v_\lambda)(\omega) = \mathbf{\Phi}(\lambda, v_\lambda)(\omega)$$

and similarly

$$v_{n+1}(\omega) = \mathtt{val}_{X \times Y} \mathbf{\Phi}_{xy}(\frac{1}{n+1}, v_n)(\omega) = \mathbf{\Phi}(\frac{1}{n+1}, v_n)(\omega)$$

The proofs are like the ones for the recursive formula in Proposition 3.20 or Proposition 4.21. However, due to the symmetric information, a direct approach is possible based only on the properties of the operator $\mathbf{\Phi}$ (compare with Proposition C.14).

Proposition 5.1
Let $f \in \mathcal{F}$, $\varepsilon > 0$ and x be a measurable map from Ω to X such that for any ω:

$$\mathbf{\Phi}_{x(\omega)y}(\lambda, f)(\omega) \geq f(\omega) - \varepsilon, \quad \forall y \in Y.$$

Then the associated stationary strategy $\sigma = \{x(\omega)\}_{\omega \in \Omega}$ satisfies:

$$\gamma_\lambda(\sigma, \tau)(\omega) \geq f(\omega) - \varepsilon/\lambda, \quad \forall \tau.$$

Proof
We consider the conditional payoff at stage m which satisfies, $\forall \tau$:

$$E_{\sigma,\tau}(\lambda g_m + (1-\lambda) f(\omega_{m+1}) | \mathcal{H}_m) = \mathbf{\Phi}_{x(\omega_m), \tau(h_m)}(\lambda, f)(\omega_m) \geq f(\omega_m) - \varepsilon$$

We multiply this inequality by $(1-\lambda)^{m-1}$, take expectation and summation from $m = 1$ to ∞. This gives:

$$\gamma_\lambda(\sigma, \tau)(\omega_1) = E^{\omega_1}_{\sigma,\tau}(\sum_{m=1}^{\infty} \lambda (1-\lambda)^{m-1} g_m) \geq f(\omega_1) - \varepsilon/\lambda$$

hence the result. ∎

5.2.2 Auxiliary game and recursive formula

We define three conditions as follows: **C1** There exists a subspace $\mathcal{F}_0 \subset \mathcal{F}$

(that contains 0) such that $\boldsymbol{\Psi}(f)$ is defined on $\mathcal{F}_0$ (i.e. each game $\mathbf{G}(f)(\omega), \omega \in \Omega$, $f \in \mathcal{F}_0$ has a value) and $\boldsymbol{\Psi}(f)$ belongs to $\mathcal{F}_0$. **C2** For all $\varepsilon > 0$, $\alpha \in [0,1]$ and

all $f \in \mathcal{F}_0$ there exists a measurable map x from $(\Omega, \mathcal{C})$ to X, such that $x(\omega)$ is ε-optimal for Player 1 in $\Gamma(\alpha, f)(\omega)$ (and similarly for Player 2). **C3** $\mathcal{F}_0$

is complete for the uniform norm $\|\cdot\|$ on Ω. Under **C1**, one can define induc-

tively $w_1 = \boldsymbol{\Phi}(1, 0), \ldots, w_{n+1} = \boldsymbol{\Phi}(\frac{1}{n+1}, w_n)$.
Under **C1** and **C3** the operator $\boldsymbol{\Phi}(\lambda, .)$, being a contraction with parameter $(1-\lambda)$ on a complete metric space has a unique fixed point, w_λ:

$$\boldsymbol{\Phi}(\lambda, w_\lambda) = w_\lambda$$

and w_λ belongs to $\mathcal{F}_0$.

Proposition 5.2
Assume **C1, C2** *and* **C3**. *Then the discounted game has a value v_λ which equals w_λ hence satisfies:*

$$v_\lambda(\omega) = \mathtt{val}_{X \times Y} \boldsymbol{\Phi}_{xy}(\lambda, v_\lambda)(\omega)$$

Moreover each player has an ε-optimal stationary strategy.

Proof
The proof follows directly from the above remarks and Proposition 5.1. ∎

Proposition 5.3
Assume **C1** *and* **C2**. *Then the finitely repeated game has a value v_n which equals w_n, hence satisfies:*

$$v_{n+1}(\omega) = \mathtt{val}_{X \times Y} \boldsymbol{\Phi}_{xy}(\frac{1}{n+1}, v_n)(\omega)$$

Moreover each player has an ε-optimal Markov strategy.

Proof
Here also the proof is a consequence of the previous remarks. We use **C2** to define inductively the strategies: given the history h_m, Player 1 uses the measurable selection $x_m(\omega_m)$, $\frac{\varepsilon}{(n-m+1)}$-optimal in the remaining $(n-m+1)$ stage game. ∎

5.2.3 Value of the auxiliary game

Proposition 5.4
Let Ω be finite and I, J be compact. Assume $g(\cdot,\omega)$ and $q(\omega'|\cdot,\omega)$ separately continuous on $I\times J$.
Then **C1, C2** *and* **C3** *hold.*

Proof
The game satisfies the hypotheses of Theorem A.11. (Condition **C3** is trivial in this case). ∎

We describe here without proofs two more general situations in which **C1, C2, C3** hold, see also Section 8.

Measurable set up
The state space $(\Omega,\mathcal{C})$ is a standard Borel space and the move sets I and J are compact metric spaces. For any $C\subset\mathcal{C}$, $g(\cdot)$ and $q(C|\cdot)$ are measurable on $I\times J\times\Omega$. Finally for any $C\subset\mathcal{C}$ and any $\omega\in\Omega$, $g(\cdot,\omega)$ and $q(C|\cdot,\omega)$ are separately continuous on $I\times J$. In this case one works with the set $\mathcal{F}$ of bounded measurable functions.

Continuous set up
Ω is a metric space and I and J are compact metric spaces. For each f in $\mathcal{F}'$, which denotes the subset of continuous functions in $\mathcal{F}$, both $g(i,j,\omega)$ and $\int_\Omega f(.)dq(.|i,j,\omega)$ are for each i, jointly continuous on $\Omega\times J$ and for each j, jointly continuous on $\Omega\times I$. Then v_λ and v_n belong to $\mathcal{F}'$.

In both cases regularity conditions on the auxiliary one shot game suffice to guarantee the existence of the value in the stochastic game.

Corollary 5.5
The above results (Propositions 5.2 and 5.3) hold for any signalling structure on the moves, as soon as the state ω_n is known at each stage n by both players.

Proof
The previous proof shows the players have stationary (resp. Markov) optimal strategies in G_λ (resp. G_n). ∎

5.3 "Big Match"

The following game played a fundamental role in the development of the theory. The payoff matrix is:

	Left	*Right*
Top	1*	0*
Bottom	0	1

As soon as player 1 plays *Top* (at stage θ), the payoff (with a *) is absorbing: the payoff at each following stage is equal to g_θ, hence basically the game is over. Otherwise the game is repeated and the stage payoff is given by the entries of the matrix.

5.3.1 Recursive structure

One has $v_1 = 1/2$ and the recursive formula for v_n is :

$$(n+1)v_{n+1} = \mathtt{val}\begin{pmatrix} n+1 & 0 \\ nv_n & 1+nv_n \end{pmatrix}$$

It follows by induction that $v_n = \frac{1}{2}, \forall n$. The only optimal strategy for Player 1 is to play *Top* with probability $\frac{1}{n+1}$ at the first stage in G_n.
The recursive formula for v_λ is :

$$v_\lambda = \mathtt{val}\begin{pmatrix} 1 & 0 \\ (1-\lambda)v_\lambda & \lambda + (1-\lambda)v_\lambda \end{pmatrix}$$

This gives $v_\lambda = \frac{1}{2}$, for all λ and the only optimal strategy of Player 1 is to play *Top* with probability $\frac{\lambda}{1+\lambda}$ (stationary) in G_λ.
The optimal strategy for player 2 is to play $(\frac{1}{2}, \frac{1}{2})$, i.i.d. in both cases.
The limits of such optimal strategies (as n goes to ∞ or λ goes to 0) are obviously not interesting: Player 1 would always play *Bottom*. Nevertheless, the next section will describe a way of dealing with an "asymptotic" behavior.

5.3.2 The game on $[0,1]$

Consider the continuous time game played between time 0 and 1 (recall Section 1.3). Player 1 chooses a stopping time θ (of playing *Top*) according to a distribution ρ and Player 2 plays *Left* , with probability $f(t)$, at time t. The payoff is given by the integral from 0 to 1 of the payoff at time t, namely:

$$R(\rho, f) = \int_0^1 r(t)dt$$

with

$$r(t) = \int_0^t f(s)\rho(ds) + (1 - \rho([0,t]))(1 - f(t))$$

The first part corresponds to the absorbing payoff accumulated from the start: with probability $\rho(ds)$ absorption occurs at time s leading to an absorbing

payoff of $f(s)$. With probability $1-\rho([0,t])$ there is no absorption yet at time t and the current payoff is $1-f(t)$ leading to the second term.
Formally we thus define a game $\wp$ where the payoff is as above, the strategies of Player 1 are the positive measures of mass less than 1 on $[0,1]$ and the strategies of Player 2 are the continuous functions from $[0,1]$ to $[0,1]$.

Proposition 5.6
$\wp$ has a value ν. Moreover:

$$\lim_{n\to\infty} v_n = \lim_{\lambda\to 0} v_\lambda = \nu.$$

ε-optimal strategies in $\wp$ induce by discretization 2ε-optimal strategies in G_n and G_λ.

Proof
The existence of a value follows from Proposition A.8. In fact it is easy to check that $\rho =$ Lebesgue measure and $f\equiv\frac{1}{2}$ are optimal and $\nu = 1/2$.
The approximation results follow easily from the construction. ∎

Remark
Note that in the finite game G_n the optimal strategy for player 1 gives *Prob*($\theta \leq m$) $\approx \frac{m}{n}$ and similarly in the discounted game *Prob*($\theta \leq m$) $\approx \sum_{\ell=1}^{m}\lambda(1-\lambda)^{\ell-1}$. In both cases, when at stage m the past play corresponds approximately to a fraction $t\in[0,1]$ of the length of the game (according to the measure on the stages used to define the payoff), the probability of stopping before m is also near t. In other words a discretization of an optimal strategy in the continuous time game is an approximate optimal strategy in large games: we compare their effect on the state space for a given fraction on the length of the game (by opposition to a stage by stage convergence).

5.3.3 The uniform approach

We first prove that "simple strategies" of Player 1 cannot guarantee more than 0.
For a stationary strategy of Player 1 this is clear: the total probability of *Top* is either 0 or 1. Hence by playing always *Left* or *Right* Player 2 will obtain a payoff 0.
Similarly, facing a Markov strategy Player 2 would play *Right* until the probability of *Top* in the future is negligible and then switch to *Left* for ever.
The same result holds for strategies generated by a finite automaton. It is defined by a finite memory set M, an updating function from $M\times\{Left, Right\}$ to $\Delta(M)$ and a move mapping from M to probabilities on $\{Top, Bottom\}$. The state of memory changes at random as a function of the current state and of the move of the opponent. To each state of memory is associated a mixed move. Note that in this framework a stationary strategy of Player 2

will generate a Markov chain on M. Let him first play $(\varepsilon, 1-\varepsilon)$ i.i.d.. When the probability of being in an irreducible class in which only *Bottom* is played is near 1, then he will switch to *Left* (without changing the ergodic decomposition). Any absorbing payoff before the change would be 0 and the payoff will be non absorbing and 0 afterwards (with a high probability).
Nevertheless the "Big Match" has a uniform value.

Proposition 5.7

$$v_\infty = \frac{1}{2}$$

Proof
It is convenient to translate the payoffs by 1/2 to

$$A^* = \begin{pmatrix} a_1^* & a_2^* \\ b_1 & b_2 \end{pmatrix} = \begin{pmatrix} 1/2^* & -1/2^* \\ -1/2 & 1/2 \end{pmatrix}$$

so that the value of the matrix A with no * is 0.
Player 2 obviously guarantee 0 by playing i.i.d. an optimal strategy in the game A.
The construction of the strategy of Player 1 relies on a map ρ from $\mathbb{R}$ to $(0,1]$ that expresses the probability of playing *Top* as a function $\rho(z)$ of a statistic z of the past play.
Explicitly, define $z_1 = 0$, $z_{m+1} = -\sum_{\ell=1}^{m} a_{j_\ell}$. At stage m Player 1 plays *Top* with probability $\sigma_m(h_m) = \rho(z_m) = x_m$.
The next lemma is crucial. A risk function R allows Player 1 to control the absorbing part of the payoff.

Lemma 5.8
Assume that for some negative function R and $|\delta| \leq 1/2$:

$$\delta\rho(z) + (1-\rho(z))R(z-\delta) \geq R(z), \qquad \forall z \tag{5.1}$$

then:

$$\gamma_n^*(\sigma,\tau) = E_{\sigma,\tau}(g_\theta \mathbf{1}_{\{\theta\leq n\}}) \geq R(0), \qquad \forall\tau, \quad \forall n.$$

Proof
Let $\delta = a_{j_m}$ then one has from (5.1), using σ as defined above

$$a_{j_m}x_m + (1-x_m)R(z_{m+1}) \geq R(z_m).$$

Note that a_{j_m} is the absorbing payoff and x_m the corresponding probability. This implies, with $g_\theta = a_{j_\theta}$:

$$E(g_\theta \mathbf{1}_{\{\theta=m\}} + \mathbf{1}_{\{\theta>m\}}R(z_{m+1})) \geq E(R(z_m)\mathbf{1}_{\{\theta\geq m\}})$$

Since R is negative, summation over $1 \leq m \leq n$ gives:

$$E_{\sigma,\tau}(g_\theta \mathbf{1}_{\theta\leq n}) \geq R(z_1) = R(0).$$

∎

Lemma 5.9
Assume ρ decreasing. $\forall \varepsilon > 0, \exists N$ such that $n \geq N$ implies:

$$E_{\sigma\tau}[\sum_{m=1}^{n+1} \mathbf{1}_{\{\theta=m\}}(\sum_{\ell=1}^{m-1} b_{j_\ell})] \geq -2n\varepsilon$$

Proof
Assume $\sum_{m=1}^{n} b_{j_m} < -n\varepsilon$, then $\sum_{m=1}^{\ell} b_{j_m} < 0$ for all $n - [n\varepsilon] \leq \ell \leq n$. In particular, since val $A = 0$, $\sum_{m=1}^{\ell} a_{j_m} \geq 0$ for all $n - [n\varepsilon] \leq \ell \leq n$. But then $z_{\ell+1} \leq 0$ hence for all such ℓ, $\sigma_{\ell+1} = \rho(z_{\ell+1}) \geq \rho(0)$. So that

$$E(\mathbf{1}_{\{\theta>n\}} | \sum_{m=1}^{n} b_{j_m} < -n\varepsilon) \leq (1-\rho(0))^{[n\varepsilon]} \leq \varepsilon$$

for $n \geq N'$ large enough . Hence for $n \geq \max(N', \varepsilon)^2$ the contribution of the N' first terms is less than ε and for the remaining ones either the average payoff exceeds $-\varepsilon$ or the surviving probability is less than ε (and the payoffs are less than one). The claim follows. ∎

Lemma 5.10
For any $M > 0$ the functions

$$\rho(z) = \frac{1}{1+\max(1, M+z)^2} \quad \textit{and} \quad R(z) = -\frac{1}{\max(1, M+z)}$$

satisfy equation (5.1).

Proof
If both $M + z$ and $M + z - \delta$ are greater than 1, (5.1) reduces to

$$\delta\frac{1}{1+(M+z)^2} + (1 - \frac{1}{1+(M+z)^2})\times\frac{-1}{(M+z-\delta)} \geq -\frac{1}{(M+z)}$$

which holds by the choice of δ.
If $M + z \geq 1 > M + z - \delta$, one has to prove

$$\delta\frac{1}{1+(M+z)^2} - (1 - \frac{1}{1+(M+z)^2}) \geq -\frac{1}{(M+z)}$$

which follows from the inequality in the previous case with δ_0 satisfying $M + z - \delta_0 = 1$.
Finally if $M + z < 1$ the inequality is

$$\delta\rho(z) + (1-\rho(z))R(z-\delta) \geq -1$$

but $R(z - \delta) \geq -1$ gives the result. ∎

Since the total payoff is of the form

$$E_{\sigma\tau}\{\sum_{m=1}^{n+1} \mathbf{1}_{\{\theta=m\}}(\sum_{\ell=1}^{m} b_{j_\ell} + (n+1-\theta)g_\theta)\}$$

the proof of Proposition 5.7 follows from the previous three Lemmas by choosing M large enough. ∎

Remarks

Note that player 1 has no optimal strategy (that would be ε optimal for any positive ε).

The same proof holds for any game of the form $\begin{pmatrix} a_1^* & a_2^* & \dots & a_m^* \\ b_1 & b_2 & \dots & b_m \end{pmatrix}$ and v_∞ is the value of the matrix without $*$.

5.4 Finite case: algebraic approach

Assume Ω, I, J finite. The fact that v_λ is a fixed point of $\mathbf{\Phi}(\lambda, \cdot)$ is again used, but this condition reduces, for a given λ, to a finite family of polynomial inequalities in finitely many variables: the collection of stationary strategies $x_\lambda(\omega)$ and $y_\lambda(\omega)$ and the vector of values $v_\lambda(\omega)$.

Explicitly one has for any $\omega\in\Omega$, the existence of a triple $x_\lambda(\omega)\in\mathbb{R}^I$, $y_\lambda(\omega)\in\mathbb{R}^J$, $v_\lambda(\omega)\in\mathbb{R}$ such that:

$$\sum\nolimits_{i\in I} x_\lambda^i(\omega) = 1, x_\lambda^i(\omega) \geq 0, \sum\nolimits_{j\in J} y_\lambda^j(\omega) = 1, y_\lambda^j(\omega) \geq 0 \qquad \forall i\in I, j\in J, \omega\in\Omega$$

$$\sum\nolimits_{i\in I} x_\lambda^i(\omega)(\lambda g(i,j,\omega) + (1-\lambda)\sum\nolimits_{\omega'\in\Omega} q(\omega'|i,j,\omega)v_\lambda(\omega')) \geq v_\lambda(\omega) \quad \forall j\in J, \forall\omega\in\Omega$$

$$\sum\nolimits_{j\in J} y_\lambda^j(\omega)(\lambda(g(i,j,\omega) + (1-\lambda)\sum\nolimits_{\omega'\in\Omega} q(\omega'|i,j,\omega)v_\lambda(\omega')) \leq v_\lambda(\omega) \quad \forall i\in I, \forall\omega\in\Omega$$

Note that this system is also polynomial in λ. Thus taking as variables $(\lambda, x_\lambda(\omega), y_\lambda(\omega), v_\lambda(\omega))$ it defines a semi algebraic set in $(0,1]\times(\mathbb{R}^I\times\mathbb{R}^J\times\mathbb{R})^{\#\Omega}$ where. From the Tarski-Seidenberg elimination theorem (Benedetti and Risler, Theorem 2.21, p. 54) it admits a semi algebraic selection above $(0,1]$ hence we have:

Proposition 5.11

Given a finite stochastic game, there exists $\lambda_0 > 0$, $M\in\mathbb{N}$, $r_m\in\mathbb{R}^\Omega, m\in\mathbb{N}$, such that the following expansions

$$v_\lambda(\omega) = \sum\nolimits_{m=0}^{\infty} r_m(\omega)\lambda^{\frac{m}{M}}, \qquad \forall\omega\in\Omega$$

hold for all $\lambda\in(0,\lambda_0)$ and all $\omega\in\Omega$.

Proof
A semi algebraic function has an expansion near 0 in Puiseux series (Foster, Theorem 8.14, p. 58) and since the payoffs are uniformly bounded all powers are non negative. ∎

The same argument implies that similar expansions exist for $x_\lambda(\omega)$ and $y_\lambda(\omega)$.

Corollary 5.12
There exists $\lambda_1 > 0$, $M \in \mathbb{N}$ and some constant C such that the following upper bound:

$$\|\frac{dv_\lambda}{d\lambda}\| \leq C\lambda^{\frac{-(1-M)}{M}}$$

holds in $(0, \lambda_1)$.

Proof
Use Proposition 5.11 and differentiate in the interior of the disc of convergence. ∎

Proposition 5.13
Given a finite stochastic game, $\lim_{\lambda\to 0} v_\lambda$ exists.

Proof
Follows clearly from the expansion in Proposition 5.11.

We also deduce:

Proposition 5.14
In a finite stochastic game $\lim_{n\to\infty} v_n$ exists and equals $\lim_{\lambda\to 0} v_\lambda$.

Proof
From Corollary 5.12 v_λ is of bounded variation and we apply Theorem C.8. ∎

5.5 Absorbing games: asymptotic approach

The purpose of the current section is to prove asymptotic results without finiteness assumptions, using an operator approach. However only the subclass of absorbing games is covered.

Definition
An **absorbing state** ω satisfies $q(\{\omega\}|i,j,\omega) = 1$ for all i, j. An **absorbing game** is a game where all states except one, ω_0, are absorbing. A basic example is the previous "Big match " (Section 3). It is thus enough to describe the game starting from ω_0 and we drop the references to this state. I and J are

compact sets and the payoff g is separately continuous on $I \times J$. Moreover, for each $C \in \mathcal{C}$, $q(C|i,j)$ is separately continuous on $I \times J$. Finally r is a bounded and measurable absorbing payoff defined on $\Omega \setminus \{\omega_0\}$. Let again $X = \Delta(I)$ and $Y = \Delta(J)$. In this setup the domain of the Shapley operator can be reduced to the payoff in the state ω_0. Hence one considers the operator on $\mathbb{R}$ defined by:

$$\mathbf{\Phi}(\alpha, w) = \mathtt{val}_{X \times Y}\{\alpha g(x,y) + (1-\alpha) E_{x,y}(\tilde{w}(\tilde{\omega}))\}$$

where $\tilde{w}$ is equal to w on the non absorbing state ω_0 and to the absorbing payoff r elsewhere. Note that the only relevant parameters are, for each (i,j), the non absorbing payoff $g(i,j)$, the probability of absorption $(1-q(\{\omega_0\}|i,j))$ and the absorbing part of the payoff $(\int_{\Omega \setminus \{\omega_0\}} r(\omega) q(d\omega|i,j))$. Thus one can assume that there are only two absorbing states, with payoff $\|g\|$ and $-\|g\|$). Clearly the conditions of Appendix C.4 are satisfied. In the current framework Proposition C.24 has the simple following form:

Corollary 5.15
Assume $w_2 > w_1$. Then:

$$\varphi^*(w_1) - \varphi^*(w_2) > w_2 - w_1.$$

Recall that $\varphi^*(w) = \lim_{\alpha \to 0} \dfrac{\mathbf{\Phi}(\alpha, w) - w}{\alpha}$ and that $|\mathbf{\Phi}(\alpha, w)| \leq \max(\|g\|, |w|)$, hence Corollary 5.15 implies:

Corollary 5.16
There exists a single number w such that:

$$w' < w \Rightarrow \varphi^*(w') > 0$$

$$w" > w \Rightarrow \varphi^*(w") < 0.$$

Note that this w satisfies $w = \mathbf{\Phi}(0, w)$ hence $\varphi(w) = \varphi^*(w)$. The main result is now:

Theorem 5.17

$$\lim_{\lambda \to 0} v_\lambda = \lim_{n \to \infty} v_n = w.$$

Proof
The proof follows from Corollary C.23 and Proposition C.17. ∎

5.6 Infinite game: the value

The main objective of this section is to prove the following fundamental result:

Theorem 5.17
Any finite stochastic game has a uniform value v_∞.
Moreover for any positive constants ε *and* λ_0, *there exist* ε*-optimal strategies that, at each stage, play optimally in some* λ *discounted game with* $0 < \lambda \leq \lambda_0$.

Proof
By Proposition 5.11, $\lim_{\lambda \to 0} v_\lambda = v$ exists. We are looking for a strategy σ which, if $\mathcal{H}_m$ denotes the $\sigma-$algebra generated by H_m and g_m is the payoff at stage m, simultaneously controls the value function:

$$E_{\sigma,\tau}\left(v(\omega_{m+1}) - v(\omega_m)|\mathcal{H}_m\right) \geq 0 \quad \forall \tau,$$

and the current payoffs for n large enough:

$$E_{\sigma,\tau}\left(\sum\nolimits_{m \leq n}(g_m - v(\omega_m))\right) \geq 0 \qquad \forall \tau.$$

The first inequality corresponds to $v(\omega_m)$ being a bounded submartingale and the second then implies $E(\overline{g}_n) \geq v(\omega_1)$.
Note that these inequalities can usually not be satisfied (see Section 3 for example). However one can allow for an error α_m in the first one (with $\sum_m \alpha_m$ small) and similarly the second inequality has to be satisfied up to some ε.
The basic idea is to define inductively a sequence of discount factors λ_m, measurable w.r.t. $\mathcal{H}_m$ and to let Player 1 play optimally at stage m in the λ_m discounted game starting from state ω_m. With such a strategy σ, for any τ of Player 2 one obtains:

$$E_{\sigma,\tau}\left(\lambda_m g_m + (1-\lambda_m)v_{\lambda_m}(\omega_{m+1})|\mathcal{H}_m\right) \geq v_{\lambda_m}(\omega_m)$$

that we write as:

$$E_{\sigma,\tau}\left(v_{\lambda_m}(\omega_{m+1}) - v_{\lambda_m}(\omega_m) + \lambda_m(g_m - v_{\lambda_m}(\omega_{m+1}))|\mathcal{H}_m\right) \geq 0. \tag{5.2}$$

A function φ is first constructed as follows. Recall from Corollary 5.12 that there exists $\overline{\lambda} > 0$ and an integrable positive function ψ on $(0, \overline{\lambda}]$ such that for any $0 < \lambda_1 < \lambda_2 \leq \overline{\lambda}$:

$$\|v_{\lambda_1} - v_{\lambda_2}\| \leq \int_{\lambda_1}^{\lambda_2} \psi(t)dt. \tag{5.3}$$

Let $A = \max_{I \times J \times \Omega}|g(i,j,\omega)|$ and $0 < \varepsilon < A$. Define

$$s(y) = \frac{12A}{\varepsilon} \int_y^{\overline{\lambda}} \frac{\psi(t)}{t} dt + \frac{1}{\sqrt{y}} \tag{5.4}$$

Note that s is strictly decreasing. We thus can introduce:

$$\varphi(\lambda) = \int_{s(\lambda)}^{\infty} s^{-1}(t)dt$$

which is finite since:

$$\int_0^{\overline{\lambda}} \int_y^{\overline{\lambda}} \frac{\psi(t)}{t}\, dt\, dy = \int_0^{\overline{\lambda}} \psi(t)dt < \infty$$

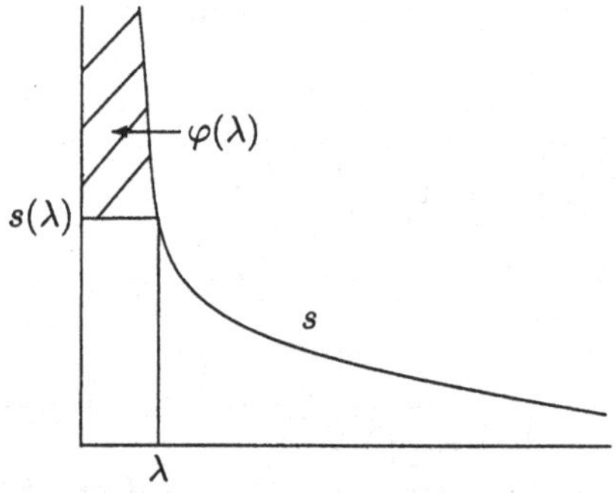

Fig. 5.1.

hence $\int_0^{\overline{\lambda}} s(y)dy =< \infty$ as well. Consider $Z_m = v_{\lambda_m}(\omega_m) - \varphi(\lambda_m)$. For λ_m small, $v_{\lambda_m}$is near v and $\varphi(\lambda_m)$ is small hence Z_m is near $v(\omega_m)$. From (5.2) one obtains:

$$E_{\sigma,\tau}(Z_{m+1} - Z_m|\mathcal{H}_m) \geq E_{\sigma,\tau}(C_1 - C_2 - C_3|\mathcal{H}_m) \tag{5.5}$$

with:

$$\begin{aligned} C_1 &= \varphi(\lambda_m) - \varphi(\lambda_{m+1}) \\ C_2 &= v_{\lambda_{m+1}}(\omega_{m+1}) - v_{\lambda_m}(\omega_{m+1}) \\ C_3 &= \lambda_m(g_m - v_{\lambda_m}(\omega_{m+1})) \end{aligned}$$

To control these quantities the sequence λ_m is explicitely defined next. Given $s_1 \geq B$ to be specified later, one introduces inductively:

$$\begin{aligned} \lambda_m &= s^{-1}(s_m) \\ s_{m+1} &= \max\{B, s_m + g_m - v_{\lambda_m}(\omega_{m+1}) + 4\varepsilon\}. \end{aligned}$$

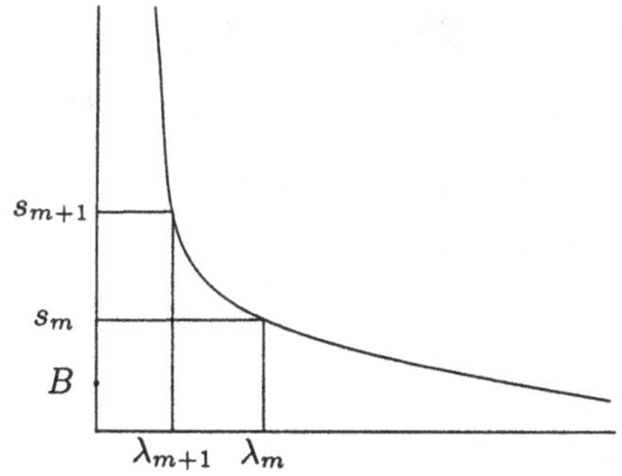

Fig. 5.2.

By construction one has

$$|s_{m+1} - s_m| \leq 6A. \tag{5.6}$$

Since, from (5.4), $s((1-\varepsilon)y) - s(y) \to \infty$ as $y \to 0$ one has, for $B \geq B_1$ large enough, using (5.6):

$$|\lambda_{m+1} - \lambda_m| \leq \frac{\varepsilon \lambda_m}{6A} \tag{5.7}$$

so that the mean value theorem gives, using (5.6) and (5.7):

$$C_1 = \varphi(\lambda_m) - \varphi(\lambda_{m+1}) \geq \lambda_m(s_{m+1} - s_m) - \varepsilon\lambda_m. \tag{5.8}$$

On the other hand, by (5.3), thanks again to (5.6) and (5.7):

$$\begin{aligned} C_2 &= v_{\lambda_{m+1}}(\omega_{m+1}) - v_{\lambda_m}(\omega_{m+1}) \leq |\int_{\lambda_m}^{\lambda_{m+1}} \psi(x)dx| \\ &\leq |\int_{\lambda_m}^{\lambda_{m+1}} \frac{\psi(x)}{x} dx| 2\lambda_m \leq \frac{\varepsilon}{6A}|s_{m+1} - s_m|\lambda_m \leq \varepsilon\lambda_m. \end{aligned} \tag{5.9}$$

Finally from the definition of s_m:

$$s_{m+1} - s_m \geq g_m - v_{\lambda_m}(\omega_{m+1}) + 4\varepsilon$$

hence

$$C_3 = \lambda_m(g_m - v_{\lambda_m}(\omega_{m+1})) \leq \lambda_m(s_{m+1} - s_m) - 4\varepsilon\lambda_m. \tag{5.10}$$

Replacing C_1, C_2, C_3, in (5.5) we obtain, by (5.8), (5.9), (5.10), the submartingale property:

$$E_{\sigma,\tau}(Z_{m+1} - Z_m|\mathcal{H}_m) \geq 2\varepsilon\lambda_m. \tag{5.11}$$

In particular one deduces :

$$E_{\sigma,\tau}^{\omega_1}(v_{\lambda_{m+1}}(\omega_{m+1})) \geq v_{\lambda_1}(\omega_1) - \varphi(\lambda_1)$$

so that for $B \geq B_2$ large enough, using (5.9):

$$E_{\sigma,\tau}(v_{\lambda_m}(\omega_{m+1})) \geq v(\omega_1) - 3\varepsilon. \tag{5.12}$$

To minorize the payoffs we use again the definition of s_m so that:

$$s_{m+1} - s_m \leq g_m - v_{\lambda_m}(\omega_{m+1}) + 4\varepsilon + 2A\mathbf{1}_{\{s_{m+1}=B\}}.$$

This implies that:

$$E^{\omega_1}_{\sigma,\tau}\left(\sum\nolimits_{m\leq n} g_m\right) \geq E^{\omega_1}_{\sigma,\tau}\left(\sum\nolimits_{m\leq n} v_{\lambda_m}(\omega_{m+1})\right) + E^{\omega_1}_{\sigma,\tau}(s_n - s_1)$$
$$-4n\varepsilon - 2AE^{\omega_1}_{\sigma,\tau}\left(\sum\nolimits_{m\leq n} \mathbf{1}_{\{s_{m+1}=B\}}\right). \tag{5.13}$$

Note also that $\mathbf{1}_{\{s_{m+1}=B\}} = \mathbf{1}_{\{\lambda_{m+1}=s^{-1}(B)\}}$ and that from (5.11) and the monotone convergence theorem:

$$2\varepsilon E^{\omega_1}_{\sigma,\tau}\left(\sum \lambda_m\right) \leq 4A$$

for $B \geq B_3$ large enough. In particular, for any $\delta > 0$:

$$E^{\omega_1}_{\sigma,\tau}(\#\{m; \lambda_m \geq \delta\}) \leq \frac{2A}{\varepsilon\delta}.$$

Finally from (5.13) one deduces that for B large enough there exists N such that $n \geq N$ implies:

$$E^{\omega_1}_{\sigma,\tau}(\overline{g}_n) \geq v(\omega_1) - \varepsilon \qquad \forall \tau.$$

■

Comments

The previous proof shows also that the strategy described has the following property:

$$E^{\omega}_{\sigma,\tau}(\liminf_{n\to\infty} \overline{g}_n) \geq v(\omega) - \varepsilon$$

Note in addition that the above strategy is at each stage a function only of the stream of past payoffs. However the results differ when this information is not available, see Exercises 9.4 and 9.5 .

5.7 Markov decision processes

5.7.1 An ergodic theorem

When no players are present, a stochastic game reduces to a Markov chain. In this section we will follow the traditional notations in this area. We denote by Ω the countable state space, its elements by $i, j, \ldots$ and the transition probability by P, P_{ij} is the probability of going from i to j. g is the payoff

on Ω.
The finite value v_n is thus given by $\frac{1}{n}(I + P + P^2 + ... + P^{n-1})g$ and the discounted value v_λ equals $(\lambda I + \lambda(1-\lambda)P + ... + \lambda(1-\lambda)^n P^n + ...)g$.
Generalized evaluations will also be considered (see Chapter 1, Subsection 1.3.5). Given a probability μ on $\mathbb{N}$ define a transition probability $\mu{*}P$ on Ω by:

$$\mu{*}P = \sum_{n=0}^{\infty} \mu(n) P^n.$$

The associated value v_μ is thus $\mu{*}P\ g$. A measure of "uniformity" of the probability μ is :

$$[[\mu]] = \mu(0) + \sum_{n=1}^{\infty} |\mu(n) - \mu(n-1)|.$$

Note that:

$$P(\mu{*}P) - \mu{*}P = (\mu{*}P)P - \mu{*}P = -\mu(0)I + \sum_{n=1}^{\infty} (\mu(n) - \mu(n-1))P^n.$$

Hence, with $\|M\| = \sup_i \sum_j |M_{ij}|$ one obtains:

$$\|P(\mu{*}P) - \mu{*}P\| = \|(\mu{*}P)P - \mu{*}P\| \leq [[\mu]]. \tag{5.14}$$

Note that whenever μ is decreasing $(\mu(n-1) \geq \mu(n))$, $[[\mu]] = 2\mu(0)$.

Theorem 5.18
There exists Q from Ω^2 to $\mathbb{R}$ such that for any family of probabilities $\{\mu^t\}_{t>0}$ on $\mathbb{N}$ satisfying:

$$\lim_{t\to 0} [[\mu^t]] = 0 \tag{5.15}$$

one has

$$lim_{t\to 0}\ \mu^t{*}P = Q.$$

Moreover Q satisfies:

$$Q_{ij} \geq 0,\ \forall i,j \qquad \sum_j Q_{ij} \leq 1, \forall i \tag{5.16}$$

$$PQ = Q \tag{5.17}$$

$$QP = Q. \tag{5.18}$$

Proof
Let Q be an accumulation point (by the diagonal procedure) of the family $\{\mu^{t_m}{*}P\}$ along some sequence $\{t_m\}$ going to 0.
(5.16) is clear and (5.17) follows from (5.14) and (5.15) since $\lim_{t_m\to 0} P(\mu^{t_m}{*}P)$ $= P \lim_{t_m\to 0}(\mu^{t_m}{*}P)$.
(5.14) also implies $QP \leq Q$ since by Fatou's Lemma and (5.15):

$$(\lim_{t_m\to 0}(\mu^{t_m}{*}P))P \leq \lim_{t_m\to 0}((\mu^{t_m}{*}P)P) = Q$$

Finally $Q_{i\ell} > \sum_j Q_{ij}P_{j\ell}$ for some couple i, ℓ implies:

$$\sum_k Q_{ik} > \sum_k \sum_j Q_{ij}P_{jk} = \sum_j \sum_k Q_{ij}P_{jk} = \sum_j Q_{ij}$$

a contradiction, so that (5.18) holds.
Let now Q' be an accumulation point as t goes to 0, of some alternative family $\{\beta^t {*} P\}$ satisfying (5.15). (5.18) gives $Q(\beta^t {*} P) = Q$ hence

$$QQ' = Q$$

as well. On the other hand (5.17) implies $(\beta^t {*} P)Q = Q$, hence by Fatou's Lemma again

$$Q'Q \leq Q$$

Dual relations starting from Q' imply $Q = Q'$. ∎

Remark
Let $n\overline{P}(n) = \sum_{m=0}^{n-1} P^m$ and $\overline{P}(\lambda) = \sum_{n=0} \lambda(1-\lambda)^n P^n$. Both limits of $\overline{P}(n)$ as n goes to ∞ or of $\overline{P}(\lambda)$ as λ goes to 0 exist and are equal to Q. One has $PQ = Q$ which is clear in a dynamic programming approach, but the meaning of $QP = Q$ is less intuitive.

Corollary 5.19
Q is the unique point in the closed convex hull of $\{P^n\}$ satisfying:

$$PQ \leq Q \qquad \text{and} \qquad QP \leq Q.$$

Note however that in the countable case the results on the asymptotics of the process do not translate to the asymptotics of the value: Let $\Omega = \mathbb{N}$ with deterministic transition $n \rightarrow n+1$ and payoff g identically 1. One has $v_n = v_\lambda = v_\infty \equiv 1$ while the only invariant measure is 0 and $Q \equiv 0$. So that $\lim_{t\to 0}\{(\mu^t {*} P)(g)\} \neq (\lim_{t\to 0} \mu^t {*} P)(g)$.

5.7.2 Finite dynamic programming

This section is devoted to the study of the one person case when Ω and I are finite.
In the equations defining the value v_λ the operator $\mathtt{val}_{\Delta(I)\times\Delta(J)}$ is replaced by $\max_I$ hence it is enough to consider pure strategies.

Proposition 5.20
There exists $\overline{\lambda}$ and a pure stationary strategy x (from Ω to I) which is optimal in G_λ for all $\lambda \in (0, \overline{\lambda}]$. Moreover v_λ has an expansion in power series of λ.

Proof
From Proposition 5.11 there exists $x_\lambda(\omega)$ optimal and algebraic with value

in I, thus it is constant and equal to $x(\omega)$ in a neighborhood of 0. Since v_λ satisfies:

$$v_\lambda(\omega) = \lambda g(x(\omega), \omega) + (1-\lambda) \sum_{w'} q(\omega'|\omega, x(\omega)) v_\lambda(\omega')$$

it can be expressed as a rational fraction of λ hence the result. ∎

Thus v_λ converges to some w and from Proposition 5.18 applied to the transition $q(\omega, x(\omega); \omega')$ one obtains $w = \lim_{n\to\infty} \overline{\gamma}_n(x)$. In addition, in the one person case the convergence of v_n is more precise:

Proposition 5.21
$n\|v_n - w\|$ is uniformly bounded.

Proof
From $\frac{v_\lambda}{\lambda} = \Psi((1-\lambda)\frac{v_\lambda}{\lambda})$ one obtains, since v_λ has an expansion in power series and Ψ is algebraic and non-expansive that

$$\frac{w}{\lambda} + v_1 = \Psi(w(\frac{1}{\lambda} - 1) + v_1)$$

with $v_\lambda = w + v_1\lambda + 0(\lambda^2)$. So that

$$nw + v_1 = \Psi((n-1)w + v_1)$$

and the results follows since $nv_n = \Psi((n-1)v_{n-1})$. ∎

5.7.3 Compact move spaces

We still consider a finite state space but we leave the algebraic aspect. The move space I is compact, the transition q from $I \times \Omega$ to $\Delta(\Omega)$ and the reward g from $I \times \Omega$ to $\mathbb{R}$ are continuous in i. A stationary strategy $x \in \chi = I^\Omega$ defines trough the transition $q(\omega'|x(\omega), \omega)$ a Markov chain P_x on the finite state space Ω and the corresponding ergodic limit (Proposition 5.18) is denoted by Q_x. Given such an x, the asymptotic payoff is defined by $\gamma(x, \omega) = \lim_{n\to\infty} \overline{\gamma}_n(x, \omega) = \lim_{\lambda\to 0} \gamma_\lambda(x, \omega) = [Q_x g(x)]_\omega$ where $g(x)$ stands for the vector $\{g(x(\omega), \omega)\}$ (and similarly $\gamma(x)$ is the vector $\{\gamma(x, \omega)\}$). Let:

$$W(\omega) = \sup_{x \in \chi} \gamma(x, \omega)$$

This is the value of the "infinite game" reduced to stationary strategies. We first prove two results using the analysis of the previous section.

Proposition 5.22
For all $\varepsilon > 0$ there exists a stationary strategy x with:

$$\gamma(x,\omega) \geq W(\omega) - \varepsilon \qquad \forall\omega.$$

Proof
For any initial state ω there exists $x_\omega \in \chi$ satisfying:

$$\gamma(x_\omega,\omega) \;\geq\; W(\omega) - \varepsilon.$$

Since Ω is finite, the collection $\chi_0 = \{x_\omega\}$ of such pure strategies defines a finite subset I_0 of I. By Proposition 5.20 here exists within the model (Ω, I_0) a strategy x^* such that:

$$\gamma(x^*) \;\geq\; \gamma(x) \quad \forall x \in \chi_0$$

hence the result. ∎

Remark
Example 4.4 in Chapter 1 shows that one cannot do better.

Recall that one has, with the new notations:

$$\mathbf{\Phi}(\alpha, f) = \max_{x\in\chi}\{\alpha g(x) + (1-\alpha)P_x f\}.$$

Proposition 5.23

$$\mathbf{\Phi}(0, W) = W$$

Proof
For any $x, y \in \chi$ the limiting payoff corresponding to σ: play y once then always x exists and is $P_y\gamma(x)$. Using again the finiteness of Ω and Proposition 5.20, there exists $z \in \chi$ that dominates σ: $P_y\gamma(x) \leq \gamma(z)$. In particular this implies $P_y\gamma(x) \leq W$ and by taking x uniformly ε-optimal, which exists by Proposition 5.22 one achieves:

$$P_y W \leq W \qquad \forall y \in \chi \tag{5.19}$$

so that:

$$\mathbf{\Phi}(0, W) = \sup_{y\in\chi} P_y W \leq W.$$

Clearly with x ε-optimal, one obtains, using Theorem 5.18:

$$W - \varepsilon \leq \gamma(x) \leq P_x\gamma(x) \leq \mathbf{\Phi}(0, W)$$

hence W satisfies:

$$\mathbf{\Phi}(0, W) = W.$$

∎

Consider now the discounted case.

Proposition 5.24

$$W = \lim_{\lambda\to 0} v_\lambda.$$

Proof
For any x, Theorem 5.18 implies:

$$\gamma(x) = Q_x\ g(x) = \lim_{\lambda\to 0} \overline{P}_x(\lambda)\ g(x) = \lim_{\lambda\to 0} \gamma_\lambda(x).$$

In particular one obtains:

$$W \le \liminf_{\lambda\to 0} v_\lambda. \tag{5.20}$$

For each λ there exists by compactness and continuity an optimal strategy x_λ. Choose a sequence λ_m going to zero and such that: i) x_{λ_m} converges to $z \in \chi$, thus $g(x_{\lambda_m})$ converges to $g(z)$ and $P_{x_{\lambda_m}}$ converges to P_z, ii) $\overline{P}_{x_{\lambda_m}}(\lambda_m)$ converges to some matrix Π, iii) v_{λ_m} converges to some w.
One obviously has, since:

$$\overline{P}_{x_{\lambda_m}}(\lambda_m) = \lambda_m I + (1-\lambda_m) P_{x_{\lambda_m}} \overline{P}_{x_{\lambda_m}}(\lambda_m)$$

and P_x commute with $\overline{P}_x(\lambda)$, that:

$$P_z \Pi = \Pi P_z \ = \Pi \tag{5.21}$$

Now

$$v_{\lambda_m} = \gamma_{\lambda_m}(x_{\lambda_m}) = \overline{P}_{x_{\lambda_m}}(\lambda_m) g(x_{\lambda_m})$$

so that:

$$w = \Pi g(z).$$

Equality (5.21) implies as well, as in Section 7.1, that:

$$Q_z \Pi \le \Pi Q_z = \Pi$$

which leads to:

$$w = \Pi g(z) = \Pi Q_z g(z) = \Pi \gamma(z) \le \Pi W$$

Now we saw above (5.19) that $P_{x_{\lambda_m}} W \le W$, hence also $\overline{P}_{x_{\lambda_m}}(\lambda_m) W \le W$ and finally

$$\Pi W \le W$$

which gives $w \le W$, for any accumulation point w. By (5.20) we obtain $w = W$. ∎

Remark
This approach does not extend directly to the study of games in the finite state case since there are usually no stationary optimal strategy for v_n.
The crucial property used here was $P_x W \le W$ that does neither extendto the zero-sum game framework.
Finally the above result does not hold for countable move space nor for the case Ω countable, see Chapter 1, Sections 1.3.1 and 1.3.3. We consider finally

the limit of the finite games values.

Proposition 5.25

$$\lim_{n\to\infty} v_n = W$$

Proof
Ω being finite, the proof follows directly from Proposition 5.31 in the next section. ∎

Corollary 5.26
The uniform value v_∞ exists and equals W. Moreover there exists an ε-optimal stationary strategy.

5.7.4 A Tauberian theorem

We consider here a general deterministic model of Markov decision pprocess with no assumptions on the state nor the move spaces. Thus we can assume that the payoff g $(0 \le g \le 1)$ is simply a function of the current state and the transition is defined by a correspondence Γ from Ω to itself. A (feasible) play at ω is a sequence $\{\omega_\ell\}$ with $w_1 = \omega$ and $\omega_{\ell+1} \in \Gamma(\omega_\ell)$. Since the player can generate any such play one deduces:

Lemma 5.27
$\limsup_{n\to\infty} v_n(\omega_\ell)$ *and* $\limsup_{\lambda\to 0} v_\lambda(\omega_\ell)$ *are decreasing (w.r.t. ℓ) on any play.*

Another general property is

Lemma 5.28
For any $\varepsilon > 0$, any initial state ω and any number of stages n, there exists a feasible play $h = \{\omega_\ell\}$ at ω and a stage L such that:

$$\left(\frac{1}{T}\right)\sum_{\ell=0}^{T-1} g(\omega_{L+\ell}) \ge v_n(\omega) - \varepsilon, \quad \text{for all } 1 \le T \le \frac{n\varepsilon}{2}.$$

Proof
Otherwise one could find on any play some first stage less than $\frac{n\varepsilon}{2}$ such that the average payoff up to that stage would be less than $v_n(\omega) - \varepsilon$. Inductively one could thus divide any play into a sequence of paths with length at most $\frac{n\varepsilon}{2}$ with a low average payoff on each path. In particular, on any play of length n the payoff would be, up to an error of at most $\varepsilon/2$, an average of quantities less than $v_n(\omega) - \varepsilon$, hence the contradiction. ∎

The discounted average g_λ is a convex combination of $\overline{g}_n$ (recall Chapter 3, Section 3.1) but a more precise formulation is needed. Define $M(\alpha, \beta; \lambda) = \lambda^2 \sum_{\alpha \le m \le \beta} (1-\lambda)^{m-1} m$. Then it is easily checked that:

Lemma 5.29
For any bounded sequence $\{a_m\}$

$$\sum_{m=1}^{n} \lambda(1-\lambda)^{m-1}a_m = \lambda^2 \sum_{m=1}^{n-1}(1-\lambda)^{m-1}m\ (\frac{\sum_{\ell=1}^{m} a_\ell}{m})+\lambda(1-\lambda)^{n-1}n\ (\frac{\sum_{\ell=1}^{n} a_\ell}{n}) \tag{5.22}$$

There exists N_0 *and* ε_0 *such that for any* $n \geq N_0$ *and any* $\varepsilon \geq \varepsilon_0$ *one has:*

$$M((1-\varepsilon)n, n; 1/n) \geq \frac{\varepsilon}{2e} \tag{5.23}$$

For any positive constant δ*, there exists* ε_0 *such that for any* $\varepsilon \geq \varepsilon_0$ *there exists* N_0 *such that* $n \geq N_0$ *implies :*

$$M(\varepsilon n, (1-\varepsilon)n; 1/n\sqrt{\varepsilon}) \geq 1-\delta \tag{5.24}$$

In particular (5.22) implies that

Proposition 5.30
For all $\varepsilon > 0$ *and all* $N > 0$*, there exists* λ_0 *such that for any* $\lambda \leq \lambda_0$ *and any state* ω*, there exists* $n \geq N$ *with*

$$v_n(\omega) \geq v_\lambda(\omega) - \varepsilon$$

and in particular

$$\limsup_{n\to\infty} v_n \geq \limsup_{\lambda\to 0} v_\lambda.$$

The main result of this section is the next one.

Proposition 5.31
If v_λ *converges uniformly to some* w*, then* v_n *converges uniformly to* w*. Reciprocally if* v_n *converges uniformly to* w*, then* v_λ *converges uniformly to* w *as well.*

Proof
I. From v_λ *to* v_n

Lemma 5.32

$$\forall\varepsilon > 0, \exists N,\ n \geq N \Rightarrow v_n \leq w + \varepsilon.$$

Proof
The proof is by contradiction. By uniform convergence choose λ with $\|v_\lambda - w\| \leq \varepsilon/8$. Let now N such that the weight under λ of the stages after $N\varepsilon/4$ in (5.22) is less than $\varepsilon/8$. Let then $n \geq N$ and ω such that $v_n(\omega) > w(\omega)+\varepsilon$. By Lemma 5.28, we find an initial state ω_l from which all averages of sequences

of payoffs of length less than $n\varepsilon/4$ are above $v_n(\omega) - \varepsilon/2 \geq w(\omega) + \varepsilon/2$. This implies that $v_\lambda(\omega_L) \geq w(\omega) + \varepsilon/2 - \varepsilon/8$ hence $w(\omega_L) \geq w(\omega) + \varepsilon/4$: a contradiction to Lemma 5.27. ∎

Similarly one has

Lemma 5.33

$$\forall \varepsilon > 0, \exists N,\ n \geq N \Rightarrow v_n \geq w - \varepsilon.$$

Proof
Let n large and ω such that $v_n(\omega) < w(\omega) - \varepsilon$. This implies that on any play at ω the average of the nth first payoffs is below $w(\omega) - \varepsilon$ hence also $(\frac{1}{T})\sum_{\ell=1}^{T} g(\omega_\ell) \leq w(\omega) - \varepsilon/2$ for any $(1-\varepsilon/2)n \leq T \leq n$. By (5.23) one can find λ $(= 1/n)$ such that the discounted evaluation on these stages in (5.22) is at least $\delta = \varepsilon/4e$. For the remaining stages, the average payoff up to m is at most $w(\omega) + \rho$ for any small ρ if m is large enough by the previous lemma ($v_m \leq w + \rho$ for $m \geq N(\rho)$). For n large enough, the discounted weight with $\lambda = 1/n$ of the stages before $N(\rho)$ is less than ρ, hence the total evaluation under λ is of the form:

$$\rho + \delta(w(\omega) - \varepsilon/2) + (1 - \delta - \rho)(w(\omega) + \rho)$$

which is less than $w(\omega) - \delta\varepsilon/4$ for ρ small enough. ∎

II. From v_n to v_λ
We first prove a consequence of the uniform convergence of v_n.

Lemma 5.34
For any $\varepsilon > 0$ small enough, there exists N such that for any $n \geq N$ and any state ω, there exists a play with

$$(\frac{1}{T})\sum_{\ell=1}^{T} g(\omega_\ell) \geq w(\omega) - \varepsilon, \quad \text{for all } \ \varepsilon n \leq T \leq (1-\varepsilon)n.$$

Proof
Let $\delta = \varepsilon^2/3$. By uniform convergence let N such that $\|v_n - w\| \leq \delta$ for $n \geq \varepsilon N$. Let $h = \{\omega_\ell\}$ be a δ optimal play for $v_n(\omega)$. We express the fact that the average payoff on any long path cannot exceed w by much: in particular for $T \leq (1-\varepsilon)\, n$ one has

$$v_{n-T}(\omega_{T+1}) \leq w(\omega_{T+1}) + \delta \leq w(\omega) + \delta$$

by lemma 5.27. This gives a lower bound on the average payoff from the start:

$$n(v_n(\omega) - \delta) \leq \sum_{\ell=1}^{T} g(\omega_\ell) + (n-T)(w(\omega) + \delta)$$

which implies

$$\sum_{\ell=1}^{T} g(\omega_\ell) \geq Tw(\omega) - 3\delta n$$

hence the result for all $T \geq \varepsilon n$. ∎

We now deduce:

Lemma 5.35

$$\forall \delta > 0, \exists \lambda_0, \lambda \leq \lambda_0 \Rightarrow v_\lambda \geq w - \delta$$

Proof

Use (5.24) and the previous Lemma to obtain the existence of N_0 such that for $n \geq N_0$ there exists a play where all the averages between εn and $(1-\varepsilon)n$ are above $w(\omega) - \delta/3$ and the weight of these stages in (5.22) is at least $(1-\delta/3)$ when $\lambda = \lambda_n = 1/n\sqrt{\varepsilon}$. It remains to note that on any play, $\lambda_{n+1} \leq \lambda \leq \lambda_n$ implies

$$\gamma_{\lambda_{n+1}}(h)/\lambda_{n+1} \leq \gamma_\lambda(h)/\lambda \leq \lambda_n \gamma_{\lambda_n}(h)$$

and the result follows for λ small enough. ∎

The last property needed is:

Lemma 5.36

$$\forall \delta > 0, \exists \lambda_0, \lambda \leq \lambda_0 \Rightarrow v_\lambda \geq w - \delta$$

Proof

Follow from the uniform convergence of v_n and Proposition 5.30. ∎

This achieves the proof of the main result, Proposition 5.31. ∎

The extension to the stochastic transition case, under suitable measurability conditions, follows the same ideas.

5.8 Comments and extensions

The model of a stochastic game, the definition of the auxiliary one shot game $\Gamma(\alpha, f)$, the use of the contraction principle, as well as the proof of existence of a value and of stationary optimal strategies are due to Shapley (1953), who considered the finite case. The undiscounted game and the problem of existence of a value were introduced by Gillette (1957) who also constructed the "Big Match". This game was finally solved by Blackwell and Ferguson (1968). The proof of a uniform value for absorbing games was achieved by Kohlberg (1974). The existence proof of a value in the discounted case was progressively extended to larger state and move spaces, e.g. Takahashi (1962), Maitra

and Parthasarathy (1970), Frid (1973), Rieder (1978), Couwenberg (1980). A quite general result is in Nowak (1985). The algebraic approach was initiated by Bewley and Kohlberg (1976a, 1976b). In the discounted finite case, the analysis of the n-person game, in particular the algebraic properties are very similar to the one presented here, see Mertens (1982) and MSZ. Bewley and Kohlberg (1978b) provided an expansion of v_n in fractional power of n up to a term $n^{\frac{1}{M}-1}$ and one cannot do better (see Exercise 8.2). The existence of the uniform value, due to Mertens and Neyman (1981), extends the previous results of Blackwell and Ferguson (1968) and Kohlberg(1974) and uses properties established by Bewley and Kohlberg (1976a). Their result is more general that the one presented here: it only requires the payoffs to be uniformly bounded, v_λ to exist and to satisfy some "bounded variation" property. In the framework of dynamic programming, conditions for equivalence of uniform convergence with general evaluation μ are in Monderer and Sorin (1993). However this property does not imply the existence of v_∞, see Lehrer and Monderer (1994) and Monderer and Sorin (1993). Stochastic games with alternative evaluation of payoffs have also been extensively studied: for example the positive case ($g_n \geq 0$ and payoff $\sum_{n=0}^{\infty} g_n$), Nowak and Raghavan (1991) and the lim sup case (payoff $\limsup_{n\to\infty} g_n$), Maitra and Sudderth (1992, 1993, 1998).
An overview of the field can be found in Parthasarathy and Stern (1977), Parthasarathy (1984) and Mertens (2001).

Among the main open problems are:
1) the existence of $\lim_{n\to\infty} v_n$ and $\lim_{\lambda\to 0} v_\lambda$ in the finite state, compact move case (Section 5) and their equality
2) alternative conditions for v_∞ to exist
3) extension of the uniform convergence property.
For the extension to games with signals, see the next Chapter 6, Section 6.4.

5.9 Exercices

5.9.1 Irreducible case

A finite stochastic game is **irreducible** if for any couple of stationary strategies the induced Markov chain on the state space is irreducible.
Prove that in this case $\lim_{n\to\infty} v_n$ exists and is independent of z. Deduce the existence of v_∞.

5.9.2 Expansion for v_n, Bewley and Kohlberg (1976b)

Consider the following absorbing game, where a * denotes an absorbing entry:

$$\begin{pmatrix} -1^* & 1 \\ 1 & 0 \end{pmatrix}$$

Define the sequence W_n by $W_1 = 1/3$ and $W_{n+1} = W_n + \frac{1}{n+3}$. Show that $nv_n - W_n$ converges and deduce that the value satisfies $v_n \approx \frac{Log(n)}{n}$. Hence there is no expansion of v_n in fractional powers of n.

5.9.3 Example of fractional power expansion

Show that the following $n \times n$ absorbing game

$$\begin{pmatrix} 1^* & \dots & 0^* & \dots & 0^* \\ \vdots & \ddots & & & \vdots \\ 0 & \dots & 1^* & \dots & 0^* \\ \vdots & & & \ddots & \vdots \\ 0 & \dots & 0 & \dots & 1^* \end{pmatrix}$$

has a discounted value $v_\lambda = (1 - \lambda^{\frac{1}{n}})(1 - \lambda)^{-1}$.

5.9.4 Generalized Big Match, Kohlberg (1974)

Consider the following absorbing game:

$$\begin{pmatrix} \omega_1 a_1^* & \omega a_2^* & \dots & \omega_J a_J^* \\ b_1 & b_2 & \dots & b_J \end{pmatrix}$$

If Player 1 plays *Top* and Player 2 uses the move j the payoff is, with probability $\omega_j > 0$, absorbing and equal to a_j. If Player 1 plays *Bottom* the payoff is non absorbing.
Given $y \in \Delta(J)$, define $\omega(y) = \sum_j \omega_j y_j$, $a^*(y) = \sum_j \omega_j y_j a_j$ and $b(y) = \sum_j y_j b_j$. Introduce the level $\alpha = \min_{y \in \Delta(J)} \max\{\frac{a^*(y)}{\omega(y)}, b(y)\}$. Clearly Player 2 can guarantee α by playing i.i.d. a strategy realizing the min.
As for Player 1, first substract α to each entry so that the new level is 0 and equals $\min_{y \in \Delta(J)} \max\{a^*(y), b(y)\}$. This new game is actually equivalent to the game obtained by replacing each entry $\omega_j a_j^*$ by $(\omega_j a_j)^*$ and where player 1 can guarantee 0 with a strategy like in Section 3. Hence $v_\infty = \alpha$.

5.9.5 Random stopping times, MSZ.

Let $\{X_m\}$ be a family of random variables on $[0, 1]$. Let Ω be the product space of the X_m with the filtration $\mathcal{F}_m = \sigma(X_1, ..., X_m)$.
θ is a **randomized stopping time** if $P_n(\omega) = Prob(\theta(\omega) \leq n)$ is $\mathcal{F}_{n-1}$ measurable, with P_n increasing in n and $0 \leq P_n \leq 1$.
Prove that for any $0 \leq t \leq 1$ and any $\varepsilon > 0$, there exists a randomized stopping time θ and a number N such that:
i) $\forall n, \quad \sum_{m=1}^{n} (P_m - P_{m-1})(X_m - t) \geq -\varepsilon$
ii) $\forall n \geq N, \quad \overline{X}_n \geq t + \varepsilon$ implies $P_n \geq 1 - \varepsilon$.
(Construct a stochastic game as in Section 3 with θ as optimal strategy).

5.9.6 Recursive games, Everett (1957)

In a **recursive game**, the payoff in each non absorbing states is 0. Denote by Ω the finite set of non absorbing states. Assume the move sets I and J compact and the transition separately continuous. The recursive operator in this case satisfies, for $f \in \mathbb{R}^\Omega$:

$$\Psi(f)(\omega) = \Phi(0, f)(\omega) = \mathtt{val}\{\rho(x, y; \omega) + \sum_\Omega p(\omega'|x, y, \omega) f(\omega')\}$$

where $\rho(x, y; \omega)$ is the absorbing part of the payoff.
a) Recall from Appendix C, Section 4.3, that $\mathcal{S}^+ = \{f; \varphi^*(f) \leq 0\}$. Introduce the set of supercritical vectors:

$$\mathcal{E}^+ = \{f \in \mathbb{R}^\Omega; \Phi(0, f) \leq f \text{ and } f(\omega) < 0 \Rightarrow \Phi(0, f)(\omega) < f(\omega)\}.$$

Show that $\mathcal{S}^+ = \mathcal{E}^+$.
b) Prove by induction on the number of active states that the set of critical vectors: $\mathcal{E} = \overline{\mathcal{E}^+} \cap \overline{\mathcal{E}^-}$ is non empty.
Note that, by Corollary C. 23, $\mathcal{E}$ contains at most one point.
(i) For $\#\Omega = 1$, show that a fixed point of Ψ (from $\mathbb{R}$ to itself) of smallest norm is in $\mathcal{E}$.
(ii) For each real α let $R(\alpha)$ be the recursive game obtained by replacing state 1 by an absorbing state with payoff α. By induction this new game with $(\#\Omega - 1)$ states has a critical vector $\Lambda(\alpha)$ and let $L(\alpha) = (\alpha, \Lambda(\alpha))$ in $\mathbb{R}^\Omega$. Define now $u(\alpha) = \Psi(L(\alpha))(1)$ (value starting from state 1 of the one shot game with terminal payoff $L(\alpha)$). Show that u (again from $\mathbb{R}$ to itself) is non-expansive and has a fixed point of minimal norm, α^*.
It remains to show that $L(\alpha^*)$ is a critical vector.
(iii) $\alpha^* > 0$. Let $\alpha = \alpha^* - \varepsilon$ with ε positive. Then $u(\alpha) > \alpha$ for ε small enough. Let $\delta > 0$ and β, δ-close to $\Lambda(\alpha)$ be a supercritical vector in the game $R(\alpha)$ (by induction). Prove that (α, β) is a supercritical vector for δ small enough.
(iv) $\alpha^* \leq 0$. Let Ω_0 denote the set of states with $L^\omega(\alpha^*) = \alpha^*$ and Ω_1 its complement in Ω. For α real, denote by $R_0(\alpha)$ the recursive game on Ω_1 where the states in Ω_0 are absorbing with payoff α and let $\Lambda_0(\alpha)$ be its critical vector. Finally let $L(\alpha)$ be the vector in $\mathbb{R}^\Omega$ with components α in Ω_0 and $\Lambda(\alpha)$ otherwise. Note that $\Psi(L_0(\alpha^*)) = L_0(\alpha^*)$.
Prove that $\alpha < \alpha^*$ implies $\Lambda_0^\omega(\alpha^*) - \Lambda_0^\omega(\alpha) < \alpha^* - \alpha$ for $\omega \in \Omega_1$. Hence for any $\varepsilon > 0$, there exists $\delta > 0$ with $\Lambda_0^\omega(\alpha) - \delta > \Lambda_0^\omega(\alpha^*) - \varepsilon$ for $\alpha^* - \varepsilon = \alpha$. Let now β be a supercritical vector δ close to $\Lambda_0(\alpha)$ in the game $R_0(\alpha)$. Deduce from $\Psi(\Lambda_0^\omega(\alpha^*) - \varepsilon)(\omega) \geq \alpha^* - \varepsilon$ on Ω_0, that $\Psi((\Lambda_0^\omega(\alpha) - \delta)\mathbf{1}_{\Omega_1}, \alpha\mathbf{1}_{\Omega_0})(\omega) \geq \alpha$, hence $\Psi(\beta, \alpha\mathbf{1}_{\Omega_0})(\omega) \geq \alpha$ as well. Show then that $(\beta, \alpha\mathbf{1}_{\Omega_0})$ is a supercritical vector.
Conclude from Corollary C. 23 that $\lim v_n = \lim v_\lambda$ exists.
c) Note that in a recursive game the limit of the average payoffs $g = \lim \overline{g}_n$ is

well defined on plays, hence one could consider the game with payoff g. Prove that is value is v and that both players have ε-optimal Markov strategies. (Recall that Example 3.2 in Chapter 1 provides a one person recursive game where the value of the limiting game differs from the limit of the values of the finite games - but the state space is countable).
For a study of the uniform approach see Chapter 6, Section 6.5.

5.9.7 Markov decision process with no signals, Rosenberg, Solan and Vieille (2000)

Consider a finite dynamic programming problem where the current state is unknown. Starting from an initial distribution π on Ω, the choice of an move in I induces a new distribution on Ω hence $\Delta(\Omega)$ is the natural "state space". We will write w_n for $v_n(\pi)$.
a) Define a sequence n_k such that w_{n_k} converges to $w = \limsup_{n\to\infty} w_n$. From Lemma 5.28 there exists a pure strategy σ_k and a stage m_k such that all average expected payoffs under σ_k, from stage m_k on and during less than $r_k = n_k\varepsilon/2$ stages exceed $v_{n_k} - \varepsilon$. Let π_k be the state at stage m_k under σ_k and $\overline{\pi}$ be an accumulation point of the family $\{\pi_k\}$. Choose a subsequence (still denoted by k going to ∞) with $\|\pi_k - \overline{\pi}\| \leq \varepsilon/4$. Let finally τ_k be σ_1 up to stage m_1 (excluded) and σ_k from then on. Show that there exists M such that for k large enough: $\gamma_n(\tau_k) \geq w - \varepsilon$, for all $M \leq n \leq m_1 + r_k$.
b) Consider an accumulation point of the sequence τ_k and deduce that v_∞ exists.

5.10 Notes

Section 2 basically follows the original approach of Shapley (1953), who considered the finite case, see also MSZ. The analysis of the "Big match" is due to Blackwell and Ferguson (1968). The construction in 3.3 follows Coulomb (1996). The algebraic approach in the finite case (Section 4) was initiated by Bewley and Kohlberg (1976a) with a slightly different perspective: they studied the value operator on the field of Puiseux series. We follow here Mertens (1982) and MSZ. The result of Section 5 is in Rosenberg and Sorin (2001) and follows the ideas of Kohlberg (1974) for the finite move case. The content of Section 6 is due to Mertens and Neyman (1981) who gives the first general existence result of the uniform value. Section 7.2 goes back to Blackwell (1962), Section 7.3 relies on Dynkin and Yushkevich (1979) and Section 7.4 is due to Lehrer and Sorin (1992).

6 Advances

6.1 Presentation

This chapter is more in the spirit of a survey. First because the material to cover is too wide but also because the tools used are more difficult and a lot of questions are still open.
Henceforth the main objectives will be to describe precisely some directions of research and in particular to underline the deep connection between incomplete information games and stochastic games.
The first two sections deal with incomplete information games. Section 2 focuses on the case of symmetric incomplete information (by opposition to private information like in Chapters 3 and 4) and shows that one is lead to stochastic games on the beliefs space. Section 3 considers the case of "absorbing signals" - meaning that once a signal is received the future payoff is determined. The analysis reduces then to the disymmetric model where each player plays only as a function of his previous own moves. A new tool is introduced in the form of an auxiliary one shot game mimicking the asymptotic behavior of the players.
The last two sections are devoted to extensions of the model of stochastic games defined in Chapter 5. First to the case where the moves are not observed then to the situation where the state is not public knowledge. The results related to the first extension concern mainly absorbing games but they already introduce a lot of new promising tools. In the latter case, the general model is far from being covered but significant advances are presented. Final comments describe further extensions.

6.2 Incomplete information: symmetric case

In the framework of Chapter 3, the beliefs of the uninformed player played the role of state variable, in particular in the recursive formula (Section 3.5). In the current case of symmetric information, these beliefs are public and the reduction to a stochastic game is not only valid in the asymptotic analysis but extends to the uniform approach.

The simpler deterministic case is considered first, then the general random case is presented. Finally the analysis is extended to stochastic games with symmetric incomplete information structure.

6.2.1 The deterministic case

K is a finite set of states of nature called parameters. I and J are the finite sets of moves. For each k in K, G^k is a two-person zero-sum game on $I \times J$. In addition **signalling functions** ℓ^k defined on $I \times J$ with value in some signal space Ω are given. We describe the associated repeated game $\Gamma(p)$ by specifying the initial information of the players on the parameter and the additional information gathered along the play on the parameter and on the previous moves. The crucial feature of the present model is that both aspects are symmetric among the players. Given an initial probability p on K, the parameter k is chosen once and for all according to p but is not transmitted to the players. At stage n, Player 1 (resp. Player 2) chooses $i_n \in I$ (resp. $j_n \in J$). The payoff at stage n is thus $r_n = G^k(i_n, j_n)$ but is not announced. Rather the players are told the "public signal" $\omega_n = \ell^k(i_n, j_n)$. The symmetric information requirement implies that the signal contains the moves: $(i,j) \neq (i',j')$ implies $\ell^k(i,j) \neq \ell^{k'}(i',j')$.

For every vector of moves (i,j), the set of feasible signals is $\ell(i,j)) = \{\ell^k(i,j), k \in K\}$ and the signal observed induces a partition of K. For any signal ω, let $K(\omega)$ denote the set of k's compatible with it and $p(\omega)$ be the corresponding conditional probability on K. A vector of moves (i,j) is **non-revealing** (at the probability p with support K) if $\omega = \ell^k(i,j)$ is independent of k: then $K(\omega) = K$ and $p(\omega) = p$. Note that if $K(\omega) \neq K$, the cardinality of the parameter space is strictly decreasing and this will allow for an induction procedure on it. The main idea is thus that the analysis of game with symmetric and deterministic information reduces to the study of an absorbing game.

Proposition 6.1
Any zero-sum repeated game with symmetric and deterministic information has a value.

Sketch of Proof
The proof follows from the reduction above and the existence of a value for zero-sum absorbing games, Section 5.6.
Explicitly define an absorbing game with standard signalling $\Gamma'(p)$ as follows:
- if (i,j) is non-revealing at p, the state p does not change, the payoff is the average $\sum_{k \in K} p^k G^k(i,j)$ and (i,j) is announced
- otherwise with probability $\sum_{\{k; \ell^k(i,j)=\omega\}} p^k$ the new state is $p(\omega)$ and is absorbing, hence the payoff is $\left(\sum_{\omega \in \ell(i,j)} \sum_{k \in K(\omega)} p^k v(p(\omega))\right)^*$ where by induction hypothesis, the game $\Gamma(p(\omega))$ has a value $v(p(\omega))$.

The state space is thus the (finite) set of posterior probabilities that can be generated through the signals starting from p.
If Player 1 can guarantee w in $\Gamma'(p)$, he can also guarantee it in $\Gamma(p)$: in fact playing an $(\varepsilon/2)$-optimal strategy in $\Gamma'(p)$ as long as an absorbing state is not reached and using then, if the signal ω is observed, a strategy guaranteeing a payoff $v(p(\omega))$ up to $\varepsilon/2$ in $\Gamma(p(\omega))$, will define altogether an ε-optimal strategy.
Explicitly given such a strategy σ and a strategy τ of Player 2, one has:

$$E_{\sigma,\tau}(\sum_{m=1}^{n} r_m) = E_{\sigma,\tau}(\sum_{m=1}^{\theta} r_m + \sum_{m=\theta+1}^{n} r_m)$$

where θ is the stopping time corresponding to the entrance in an absorbing state in $\Gamma'(p)$. By the choice of σ after θ in the reduced game, there exists N such that if the number of remaining stages is large enough, $(n-\theta) \geq N$, one has:

$$E_{\sigma,\tau}(\sum_{m=\theta+1}^{n} r_m | \mathcal{H}_\theta) \geq (n-\theta)(v(p(\omega)) - \varepsilon/2)$$

So that if C is a bound on the payoffs one obtains:

$$E_{\sigma,\tau}(\sum_{m=1}^{n} r_m) \geq E_{\sigma,\tau}(\sum_{m=1}^{\theta} r_m + (n-\theta)v(p(\omega)) - n\varepsilon/2 - NC$$

We now use the fact that σ guarantee up to $\varepsilon/2$, w in $\Gamma'(p)$. Hence for n large enough:

$$E_{\sigma,\tau}\left(\sum_{m=1}^{\theta} r_m + (n-\theta)v(p(\omega))\right) \geq nw - n\varepsilon/2$$

and the result follows. ∎

6.2.2 Random case

We consider here the random case where for each (k,i,j), $\ell^k(i,j)$ is a probability on a finite set of signals Ω. The symmetric information hypothesis requires that if (i,j) differs from (i',j'), any $\ell^k(i,j)$ and $\ell^{k'}(i',j')$ have disjoint supports. The play of the game is like in the deterministic case, but one cannot, in case of a revealing profile, start an induction on the size of the support of the conditional probabilities. However a similar notion of revelation will be useful.
Let $\tilde{q}(p,i,j)$ be the distribution of the posterior probability on K, when the prior is p and the moves are (i,j). Explicitely let $\ell^p = \sum_{k\in K} p^k \ell^k$ and define a function q from $\Delta(K)\times I\times J\times\Omega$ to $\Delta(K)$ satisfying

$$\ell^p(i,j)(\omega) q^k(p,i,j,\omega) = p^k \ell^k(i,j)(\omega).$$

For each $(p,i,j) \in \Delta(K) \times A$, $\tilde{q}(p,i,j)$ has the following distribution:

$$Prob(\tilde{q}(p,i,j) = q(p,i,j,\omega)) = \ell^p(i,j)(\omega) = Prob(\omega|p,i,j).$$

In words $q(p,i,j,\omega)$ is the posterior distribution on K given the signal ω and $\ell^p(i,j)(\omega)$ is its probability. The subset of $I\times J$ for which $\tilde{q}(p,i,j)$ is the constant p consists of the **non-revealing** entries at p: the signal ω is non informative and the posterior does not change.

Given any real function f defined on $\Delta(K)$, introduce the following absorbing game with payoff:

$$D(f)(p)(i,j) = \begin{cases} \sum_k p^k G^k(i,j) & \text{if } (i,j) \text{ is non-revealing at } p, \\ \Big(E(f(\tilde{q}(p,i,j))\Big)^* & \text{otherwise} \end{cases}$$

and denote by $T(f)(p)$ its value. The main result is then:

Theorem 6.2
The mapping $T: f \mapsto T(f)$ has a unique fixed point v in the set of continuous functions on $\Delta(K)$.
$v(p)$ is the value of $\Gamma(p)$.

Sketch of Proof
The basic steps of the proof are as follows.
Assume by induction the result true on the boundary ∂ of $\Delta(K)$: there exists there a continuous function w with $T(w) = w$.

Lemma 6.3
Let u be continuous on $\Delta(K)$ and equal to w on the boundary. Then $T(u)$ is continuous on $\Delta(K)$ and equals w on the boundary.

The difficulty is the continuity at the boundary which follows from the fact that the value of an absorbing game is not changed when a non absorbing entry is replaced by an absorbing one with payoff equal to the value.
The next result is similar to the argument sketched in the previous section:

Lemma 6.4
If Player 1 can guarantee $u(p)$ in $\Gamma(p)$, $\forall p \in \Delta(K)$, he can also guarantee $T(u)(p)$.

We now generate monotonic sequences (recall the construction in 4.5.2):
Player 1 can guarantee $u_0(p) = max_{r\in\partial}\{w(r) - \|G\|\|r-p\|_1\}$, with $\|G\| = max_{k,i,j}|G^k(i,j)|$. Hence by induction he can also guarantee $u_{n+1} = \max\{u_n, T(u_n)\}$. This defines an increasing sequence of continuous functions on $\Delta(K)$, equal to w on the boundary ∂ and converging to some $\underline{u}$.
$\underline{u}$ is lower semi continuous, $\underline{u} = w$ on ∂, $\underline{u} \geq T(\underline{u})$ (since T is non expansive, hence continuous) and Player 1 can guarantee $\underline{u}$.

One obviously get dual results for Player 2 with a function $\overline{u}$, hence in particular $\underline{u} \leq \overline{u}$. It thus remains to show (recall Proposition 4.15):

Lemma 6.5

$$\underline{u} \geq \overline{u}$$

Proof
By contradiction let ρ in $\Delta(K)$ be an extreme point of the convex hull of the compact set where $\overline{u} - \underline{u}$ (which is u.s.c.) is maximal and equal to $\delta > 0$. Since ρ is the expectation of $\tilde{q}(\rho, i, j)$, for any (i, j) revealing at ρ one obtains by the extremality condition:

$$E((\overline{u} - \underline{u})(\tilde{q}(\rho, i, j))) < \delta$$

On the other hand, for (i, j) non-revealing at ρ the payoffs are the same in $D(\overline{u}, \rho)$ and $D(\underline{u}, \rho)$, hence one obtains for all (i, j):

$$D(\overline{u}, \rho)(i, j) - D(\underline{u}, \rho)(i, j) \ < \delta$$

so that $T(\overline{u}, \rho) - T(\underline{u}, \rho) < \delta$ and a fortiori $(\overline{u} - \underline{u})(\rho) < \delta$, a contradiction. ∎

6.2.3 Extensions

In this section an alternative proof is also obtained by induction but without relying on monotonic maps and moreover extending to non zero-sum games. In the previous construction, ε-optimal strategies at stage m were only functions of the posterior at that stage. In the present set up their construction will take into account in addition the number of stages where this value has changed. More precisely, the finiteness assumption on $I \times J$ implies that for any positive ε and any strategy pair, there is a finite number of jumps, say M, after which, with probability greater than ε, the martingale of posterior distributions will be within ε of the boundary, hence the possibility of an induction analysis on the cardinality on K.
Explicitly the strategies will be constructed as follows: at the Mth jump, choose in the boundary of $\Delta(K)$ a closest point p_* to the current value p of the martingale and play optimally in $\Gamma(p_*)$ from this stage on. This defines a payoff $e(M, p)$. Inductively payoffs $e(m, p)$ are defined on $\Delta(K)$ after m jumps ($m \leq M$). After $m - 1$ jumps, the players play at p in the stochastic game where the payoff is the expected stage payoff if the posterior does not change and is, after a jump, absorbing and equal to $e(m, p')$ where p' is the current posterior.
Hence the state space is a product $\Delta(K) \times \{1, 2, \ldots, M\}$. If there is no splitting neither p nor the counter m changes. Otherwise p evolves in the simplex and the counter increases by one.
The formal construction is as follows. Given $\varepsilon > 0$, ε-optimal strategies in

$\Gamma(p)$ will be defined. Using the Lipschitz aspect of the payoffs w.r.t. p and the induction hypothesis, it is enough to deal with p at a distance greater than $\varepsilon/2$ from the boundary of $\Delta(K)$, say $p \in \Delta'$.
From the the finiteness of $I \times J$ we deduce the existence of $\eta > 0$ such that for all p in Δ' and for all (i, j) revealing at p:

$$E(\sum_k (\tilde{q}^k(p, i, j) - p^k)^2) > \eta \tag{6.1}$$

meaning that as long as p is not near the boundary, a revealing profile (i, j) induces a variance of the posteriors uniformly bounded below by a positive number. We call such element (p, i, j) a jump. Since the sum of the per stage variation of the martingale of posteriors $\{p_n\}$, evaluated in L^2 norm is bounded, namely for each k (Lemma 3.4) :

$$\sum_{n=1}^{\infty} (p_{n+1}^k - p_n^k)^2 \leq 1$$

there exists an integer M such that the probability of the set of paths where more than M jumps occur before reaching the boundary within $\varepsilon/2$ is less than $\varepsilon/2$.
Introduce now a new state space as $\bar{K} = \Delta(K) \times \{0, 1, \cdots, M\}$ and define inductively a mapping α on $\bar{K}$ as follows:
$\sigma(M, p)$ is an $(\varepsilon/2)$-optimal strategy with payoff $\alpha(M, p)$ in the game $\Gamma(p)$ for $p \in \Delta \setminus \Delta'$ (which exists by the induction hypothesis on the number of elements in the support of p and the above remark). $\sigma(M, p)$ is arbitrarily defined for $p \in \Delta'$ and $\alpha(M, p)$ is the vector 0 there.
For $\ell = 0, 1, \cdots, M - 1$ and p near the boundary, namely $p \in \Delta \setminus \Delta'$, let $\alpha(\ell, p) = \alpha(M, p)$. Now for $\ell = 0, 1, \cdots, M-1$ and $p \in \Delta'$ define by backward procedure a game with absorbing payoffs $G'(\ell)(p)$ played on $I \times J$ by:

$$G'(\ell)(p)(i, j) = \begin{cases} \sum_k p^k G^k(i, j) & \text{if } (i, j) \text{ is non-revealing at } p, \\ \{E(\alpha(\ell + 1, \tilde{q}(p, i, j)))\}^* & \text{otherwise.} \end{cases}$$

$G'(\ell)(p)$ is an absorbing game with standard signalling with value $\alpha(\ell, p)$ and ε-optimal strategy $\sigma(\ell, p)$ hence the induction is well defined.
Let us prove that Player 1 can guarantee $\alpha(0, p)$ up to ε.
On the space of plays define W_ℓ to be the stopping time corresponding to the ℓ-th time a revealing entry is played (the ℓ-th jump), $\ell = 1, \cdots, M$ and θ to be the entrance time in $\Delta(K) \setminus \Delta'$. Let $T_\ell = \min\,(W_\ell, \theta)$.
A strategy σ^* in $\Gamma(p)$ is then constructed as follows:
σ^* coincides with $\sigma(0, p)$ until time T_1. Inductively given the past history σ^* follows $\sigma(\ell, p(\ell))$, from time $T_\ell + 1$ until time $T_{\ell+1}$, $\ell = 1, \cdots, M$, where $p(\ell)$ is the posterior distribution on K given the past history h_{T_ℓ}. More precisely for every subsequent history h, $\sigma^*(h_{T_\ell}, h) = \sigma(\ell, p(\ell))\ (h)$.
Consider finally a couple of strategies and the corresponding random path of

the martingale of posterior distribution. If the number of jumps is less than M or if the boundary is reached, the previous computations apply and the lower bound on the payoffs is satisfied. Since the complementary event has probability less than $\varepsilon/2$ this ends the proof. ∎

In fact the previous construction applies to a more general setting. Assume that rather than dealing with repeated games G^k, each of them is actually a stochastic game, played on some state space Ξ. k will refer to the uncertainty parameter while ξ will be the stochastic state parameter.
The game evolves as follows: an initial public lottery p on K selects k and then G^k is played starting from ξ_1 which is publicly known. After each stage $m \geq 1$ a public random signal ω_m is announced, which reveals the profile of moves (i_m, j_m) and the new state parameter ξ_{m+1}. The distribution of ω_m depends upon k, ξ_m and (i_m, j_m). As previously the signals induce a publicly known martingale $\tilde{p}$ of posterior distribution on K and one defines non-revealing profiles at (p, ξ) as those for which $\tilde{p} = p$. If $\tilde{\xi}$ denotes the new state, the couple of parameters $\tilde{p}, \tilde{\xi}$ is a random variable on $\Delta(K) \times \Xi$.
The family of auxiliary games is now defined on $\{1, ...M\} \times \Delta(K) \times \Xi$ by the payoff:

$$G'(\ell, p, \xi)(i, j) = \begin{cases} \sum_k p^k G^k(i, j) & \text{if } (i, j) \text{ is non-revealing at } (p, \xi), \\ & \text{and the new state is } \tilde{\xi} \\ \{E(\alpha(\ell + 1, \tilde{q}, \tilde{\xi})\}^* & \text{if } (i, j) \text{ is revealing at } (p, \xi) \end{cases}$$

where as in Section 3.2 α is constructed inductively as value of the new game. This defines a new stochastic game where absorbing states have been added. Note that if the initial games G^k are absorbing, the auxiliary game is also. Thus one obtains:

Theorem 6.6
Any finite stochastic game with incomplete symmetric information has a value.

6.2.4 Comments

The main conclusion is that as long as the information is symmetric its evolution is similar to the state process in a stochastic game. The typical incomplete information features occur only with differential information like in Chapters 3 and 4.
It is also worth mentioning that the analysis in 6.2.3. extends to the construction of equilibrium strategies in the non zero-sum case.
Two more technical remarks are:
The new state space is uncountable even when the parameter space is finite however the state process is much simpler than in a general stochastic games: namely it is a martingale which allows to reduce the analysis to absorbing

games. Note that the results in 6.2.1. and 6.2.2. only rely on the existence of a value for absorbing games (and not general stochastic games).
The analysis extends, under regularity hypothesises, to a countable or measurable signal's space. However the fact that $I\times J$ is finite is crucial in the proof to get the "minimal amount of splitting" η in case of a revealing profile.

6.3 Incomplete information games with no signals

6.3.1 Presentation

This section is devoted to the simplest model where no public information holds, hence there is no elementary state variable. The analysis is done through an auxiliary one shot game whose strategies are basically mimicking an exhaustive class of strategies in the original repeated game.
The game is defined by a finite collection of $I\times J$ matrices G^k, $k\in K$. The parameter k is chosen at random according to a probability p on K, the payoff are determined G^k but none of the players is informed. However signalling matrices H^k transmit after each stage n, given the moves (i_n, j_n), the signal $H^k_{i_n,j_n}$ to both players. The crucial condition is that the signal is either noninformative ("0") or completely revealing. In the later case both players known the true game and one can assume the payoff to be absorbing and equal to the value $\mathtt{val}G^k$ from then on. It follows that it is enough to define strategies on histories where only "0" occurred, hence the name of the game.
For each player, a strategy reduces to a function of his previous moves.
First an auxiliary game $\overline{\mathcal{G}}$ is introduced and the existence of a value is proved. Then it is shown that the minmax of the original game is $\mathtt{val}\overline{\mathcal{G}}$. In particular the disymmetry of $\overline{\mathcal{G}}$ implies that there exist games without value.
A last part is then devoted to the asymptotic analysis, done through a sequence of auxiliary games.
It will be convenient to assume each game normalized ($\mathtt{val}G^k = 0$), by substracting $\sum_k p^k \mathtt{val}G^k$, and to multiply each payoff matrix G^k by p^k so that the expected payoff is simply the sum of the conditional payoffs.

6.3.2 An auxiliary game

The game $\overline{\mathcal{G}}$ is defined by strategy sets X^* and Y and payoff function F as follows.

$$X = \oplus_{I'\subset I}\Delta(I')\times \mathbb{N}^{I\setminus I'}, \quad X^* = \oplus_{I'\subset I}\Delta(I')\times \mathbb{N}^{I\setminus I'}\times I'$$

$$Y = \oplus_{J'\subset J}\Delta(J')\times \mathbb{N}^{J\setminus J'}$$

and note the disymmetry. Given $x\in X^*$, I^x will denote the corresponding subset I', α^x will be the first component, c^x the second and i^x the last one.

Similar notions for y are J^y, β^y and d^y. The heuristic interpretation of x as a strategy of Player 1 in the original infinitely repeated game G_∞ is: play α^x i.i.d. except for $c(x) = \sum_{i\in I\setminus I'} c_i^x$ stages, uniformly distributed before some large stage N_0, at which each exceptional move i is played c_i^x times. After N_0 the move i^x is played (for y one still plays β^y i.i.d.).
To specify the payoff in $\overline{\mathcal{G}}$ that should correspond to the asymptotic payoff in G_∞ one defines first A^k as the set of non revealing moves at k:

$$A^k = \{(i,j)\in I\times J; H^k_{i,j} = 0\}$$

and its sections:

$$A^k_i = \{j\in J; (i,j)\in A^k\}, \quad A^k_j = \{i\in I; (i,j)\in A^k\}.$$

The payoff is then of the form $F(x,y) = \sum_{k\in K} F^k(x,y)$ with:

$$F^k(x,y) = \rho^k(x,y) f^k(x,y)$$

where $\rho^k(x,y)$ stands for the probability that the game k will not be revealed under x and y:

$$\rho^k(x,y) = \mathbf{1}_{\{I^x\times J^y\subset A^k\}} \prod\nolimits_j \alpha^x(A^k_j)^{d^y_j} \prod\nolimits_i \beta^y(A^k_i)^{c^x_i}$$

and $f^k(x,y)$ represents the conditional asymptotic payoff:

$$f^k(x,y) = i^x G^k \beta^y.$$

Define $X_0^* = \{x\in X; \alpha^x_i > 0, \forall i\in I^x\}$ and similarly $Y_0 = \{y\in Y; \beta^y_j > 0, \forall j\in J^y\}$.

Proposition 6.7
The game $\overline{\mathcal{G}}$ has a value and both players have ε-optimal strategies with finite support on X_0^* and Y_0.

Proof
First define a compact topology on X^* and Y. Introduce the compactification $\overline{\mathbb{N}} = \mathbb{N}\cup\{+\infty\}$ and $\overline{X}^* = \oplus_{I'\subset I}\Delta(I')\times\overline{\mathbb{N}}^{I\setminus I'}\times I'$. The map ι from $\overline{X}^*$ to X^* sends the exceptional moves played infinitely many times to the support of α, with frequency 0. Formally:

$$\iota(\overline{x}) = x \quad \text{is defined by} \begin{cases} I^x = I^{\overline{x}}\cup\{i; c^{\overline{x}}_i = +\infty\} & \\ \alpha^x_i = \alpha^{\overline{x}}_i & \text{for } i\in I^{\overline{x}} \\ \quad\;\; = 0 & \text{for } i\in I^x\setminus I^{\overline{x}} \\ c^x_i = c^{\overline{x}}_i & \text{for } i\notin I^x \\ i^x = i^{\overline{x}} & \end{cases}$$

Note that $F(\overline{x}, y) = F(\iota(\overline{x}), y)$. The strongest topology on X^* for which ι is continuous makes it compact. A similar construction holds for Y. For $y \in Y_0$, $F(\cdot, y)$ is continuous on X^* and for $x \in X_0^*$, $F(x, \cdot)$ is continuous on Y. Proposition A. 10 implies the existence of a value v^+on $\Delta(X^*) \times \Delta_f(Y_0)$ and v^- on $\Delta_f(X_0^*) \times \Delta(Y)$. Finally since for any $x \in X^*$ there is a sequence x_m in X_0^* such that $\lim_{m \to \infty} F(x_m, y) = F(x, y)$ for any $y \in Y$ and similarly for any $y \in Y$ there is a sequence y_m in Y_0 such that $\lim_{m \to \infty} F(x, y_m) = F(x, y)$ for any $x \in X^*$, one obtains $v^+ = v^-$ and the result follows. ∎

6.3.3 Minmax and maxmin

Theorem 6.8

$$\min \max G_\infty = \mathtt{val}\overline{\mathcal{G}}$$

Sketch of the proof
The proof contains two parts corresponding to the two conditions in the definition of the min max (Section 3.1).
Player 2 can guarantee $\mathtt{val}\overline{\mathcal{G}}$
Consider Υ, an ε-optimal strategy of Player 2 in $\overline{\mathcal{G}}$ with finite support in Y_0. Let $\delta > 0$ minorize any frequency appearing in β^y and d majorize all the components of d^y, for all y in the support of Υ. For N large enough define a strategy $\tau = \tau(N)$ of Player 2 in G_∞ as follows. Select some y according to Υ then play $\tau(y, N)$, namely: generate $d(y) = \sum_{j \notin J^y} d_j^y$ uniformly distributed random stages on $[1, N]$ where the corresponding exceptional moves $j \notin J^y$ are played; at all other stages use β^y i.i.d. Let ζ be a pure strategy of Player 1 in G_∞, given by a point in I^∞.
We first approximate the probability $Q^k(\zeta, \tau(y, N))$ that the game is not revealed before stage N, given k. For r large enough represent ζ as a point $x = x(\zeta, N, r)$ in X by defining as exceptional the moves played less than r times before stage N and normalizing the remaining ones. Formally let $m_i = \#\{n; \zeta_n = i, 1 \leq n \leq N\}$, $I^x = \{i; m_i \geq r\}$, $\alpha_i^x = \frac{m_i}{\sum_{\ell \in I^x} m_\ell}$ for $i \in I^x$, $c_i^x = m_i$ if $i \notin I^x$. Then the next evaluation holds.

Lemma 6.9
Given $\eta > 0$ here exists $N_0 = N_0(\eta, \delta, d)$ and $r = r(\eta, \delta, d)$ such that $N \geq N_0$ implies:

$$|Q^k(\zeta, \tau(y, N)) - \rho^k(x(\zeta, N, r), y)| \leq \eta.$$

Consider now the expected payoff at some stage $n \geq N$. Let ℓ_i^n denotes the number of times move i is played between stages N and n and $t_i^n = m_i + \ell_i^n = \#\{r; \zeta_r = i, 1 \leq r \leq n\}$. Then one has:

$$\gamma_n(\zeta,\tau) \leq E_\tau\left(\sum\nolimits_k Q^k \prod_{i\neq\zeta_n} \beta^y(A_i^k)^{\ell_i^n} \mathbf{1}_{\{\{\zeta_n\}\times J^y \subset A^k\}} \zeta_n G^k \beta^y\right) + \|G\| e_n$$

where $\sum_n e_n \leq 1/\delta$ since if $\zeta_n = i$ and $\{i\}\times J^y \not\subset A^k$ the number of times move i will occured before revelation is in expectation less than $1/\delta$. Using the previous evaluation of Q^k one obtains:

$$\gamma_n(\zeta,\tau) \leq E_\tau\left(\sum\nolimits_k \mathbf{1}_{\{I^x\times J^y A \subset^k\}} \prod_j \alpha^x(A_j^k)^{d_j^y}\right.$$

$$\left.\prod_{\substack{i\neq\zeta_n\\ i\notin I^x}} \beta^y(A_i^k)^{t_i^n} \mathbf{1}_{\{\{\zeta_n\}\times J^y \subset A^k\}} \zeta_n G^k \beta^y\right) + \|G\|(e_n + \eta).$$

It remains to notice that the term under the expectation operator is the payoff $F(x',y)$ in $\overline{\mathcal{G}}$ where the strategy x' is defined starting from $x(\zeta,N,r)$ introduced above, by: $I^{x'} = I^x \cup \{\zeta_n\}$, $\alpha_i^x = \alpha_i^{x'}$ for $i \in I^x$, $\alpha_{\zeta_n}^{x'} = 0$ if $\zeta_n \notin I^x$, $c^{x'} = t_i^n$ and finally $i^x = \zeta_n$. By the choice of τ the expectation is thus less than $F(x',\Upsilon)$ so that:

$$n\overline{\gamma}_n(\zeta,\tau) \leq n\mathtt{val}\overline{\mathcal{G}} + \varepsilon + \|G\|(\#I/\delta + n\eta + N)$$

and the result follows.

Player 1 can defend $\mathtt{val}\overline{\mathcal{G}}$

This part is more difficult and only a description of the main steps will be given. The starting point is an ε-optimal strategy Ξ of Player 1 in the auxiliary game $\overline{\mathcal{G}}$ with finite support in X_0. Given τ, strategy of Player 2 in G_∞, Player 1 can wait long enough to represent it as a distribution on Y. First the exceptional moves will be played in finite time, then the asymptotic behavior will be approximated by a distribution on finite times which will allow to implement Ξ as a strategy of Player 1 in G_∞.

Some notations are needed. For simplicity, write Ω for the set J^∞ of pure strategies of Player 2 in G_∞. Given $\omega \in \Omega$, $J(\omega)$ denotes its support: the set of moves played infinitely often on ω. For the exceptional moves, $j \notin J(\omega)$, $d_j(w)$ counts their number of occurrence. Ξ defines γ and c (similar to δ and d above).

Given some integer N and a vector $\Theta = \{\theta_{it}; i \in I, 1 \leq t \leq c\}$ where all components are distinct integers greater than N, introduce a strategy $\sigma = \sigma(\Theta,N)$ as follows: choose x according to Ξ then use $\sigma(x,\Theta,N)$ defined as : α^x i.i.d. up to stage N, then play the move i^x except at each stage θ_{it}, $i \notin I^x$, $1 \leq t \leq c_i^x$ where i is played.

A first result proves that, given $\eta > 0$, one can choose N (as a function of η, γ and τ) such that the following set $\Omega_0 = \{\omega \in \Omega$; the probability under $(k,\sigma(x,\Theta,N),\omega)$ that the game is not revealed before N is within η of

$\mathbf{1}_{\{I^x \times J(\omega) \subset A^k\}} \prod_j \alpha^x(A_j^k)^{d_j(\omega)}\}$ satisfies $\tau(\Omega_0) \leq \eta$.
One can furthermore find N' large enough so that on Ω_0 the probability that the game is not revealed between N and $n > N'$ given $(k, \sigma(x, \Theta, N), \omega)$ and the fact that it was not revealed at stage N, is given by

$$\mathbf{1}_{\{J(\omega) \subset A_{ix}^k\}} \prod_i \prod_{t=1}^{c_i^x} \mathbf{1}_{\{\omega(\theta_{it}) \in A_i^k\}} \mathbf{1}_{\{\omega(n) \in J(\omega)\}}.$$

It follows that for n large enough one obtain a minoration:

$$\gamma_n(\sigma(\Theta, N), \tau) \geq E_\tau[\varphi\{(\omega(\theta))_{\theta \in \Theta}, \omega(n), \omega\}] - O(\eta)$$

where φ is given by:

$$\varphi\{(\omega(\theta))_{\theta \in \Theta}, \omega(n), \omega\} = E_\Xi \Bigg[\sum\nolimits_k \mathbf{1}_{\{I^x \times J(\omega) \subset A^k\}} \prod_j \alpha^x (A_j^k)^{d_j(\omega)} \mathbf{1}_{\{J(\omega) \subset A_{ix}^k\}}$$
$$\prod_i \prod_{t=1}^{c_i^x} \mathbf{1}_{\{\omega(\theta_{it}) \in A_i^k \cap J(\omega)\}} G_{ix, \omega(n)}^k \mathbf{1}_{\{\omega(n) \in J(\omega)\}} \Bigg].$$

Let $\widetilde{\omega}(n)$ denotes the random move of Player 2 at stage n (in $L_\infty(\Omega, \tau)^{\#J}$) and writes $\Phi[(\widetilde{\omega}(\theta))_{\theta \in \Theta}, \widetilde{\omega}(n)] = E_\tau[\varphi\{(\omega(\theta))_{\theta \in \Theta}, \omega(n), \omega\}]$. Let F be the closed convex hull of the set of $\sigma(L_\infty, L_1)$ limit points of $\{\widetilde{\omega}(n)\}$. Finally define Φ^* from $F \times F$ by $\Phi^*(f, g) = \Phi[\{f\}, g]$ where $\{f\}$ denotes the Θ-vector f. Theorem A.11 shows that the game Φ^* has a value ν; let κ be an optimal strategy for Player 1 with finite support on F and note that by linearity Player 2 has an optimal pure strategy g^*. Obviously if $f = g$, the behavior of Player 2 at exceptional moves (on the support of Θ) and at stage n is the same hence $\Phi^*(f, f) = \Phi[\{f\}, g]$ appears as a feasible payoff in the auxiliary game $\overline{\mathcal{G}}$. By the choice of Ξ one obtains $\Phi^*(f, f) \geq \mathtt{val}\overline{\mathcal{G}}$, for all f in F, which finally implies: $\nu \geq \Phi^*(g^*, g^*) \geq \mathtt{val}\overline{\mathcal{G}}$.
It remains to see that any f in F is a limit in probability of some sequence of convex combination of the family $\{\widetilde{\omega}(n)\}$. This allows to define a distribution $\mu = \mu^f$ on times after N such that:

$$P(\|f - \sum\nolimits_n \mu(n) \widetilde{\omega}(n)\| \geq \eta) \leq \eta$$

and so that by choosing the components of Θ i.i.d. according to μ one obtains, Φ being multilinear on $\omega(\theta)$

$$E_{\mu^f} \Phi[(\widetilde{\omega}(\theta))_{\theta \in \Theta}, \widetilde{\omega}(n)] \geq \Phi[\{f\}, \widetilde{\omega}(n)] - O(\eta).$$

The result follows by defining the choice of Θ by Player 1 as: first select f according to κ then choose the exceptional times according to μ^f.

∎

Examples of games where $\mathtt{val}\ \overline{\mathcal{G}} \neq \mathtt{val}\ \underline{\mathcal{G}}$ hence min max and max min differ can be found in Mertens and Zamir (1976b) and Waternaux (1983a, 1983b).

6.3.4 Asymptotic approach

The analysis relies on an auxiliary game of the same kind as in the previous section, however the original game being essentially finite one cannot use the same approximation at infinity. The basic idea is to consider G_n as being played on L large blocks on which the strategies can be approximated as above. One consider then the family of such games $\mathcal{G}_L$ as L goes to ∞. Formally $\mathcal{G}_L$ is defined by strategy sets X^L and Y^L where X and Y are as before and payoff function F_L given for $x = \{x(\ell)\}$ and $y = \{y(\ell)\}$ by:

$$F_L(x,y) = \sum\nolimits_{k\in K} \frac{1}{L} \sum\nolimits_{\ell=1}^{L} \left[\prod_{m=0}^{\ell-1} \rho^k(x(m), y(m)) \right] f^k(x(\ell, y(\ell))$$

where as above $\rho^k(x,y)$ stands for the probability that the game k will not be revealed under $x \in X$ and $y \in Y$ and $f^k(x,y)$ describes the payoff on $X \times Y$:

$$f^k(x,y) = \alpha^x G^k \beta^y.$$

As in 3.2 one shows that the game $\mathcal{G}_L$ has a value w_L and that the players have ε-optimal strategies with finite support on X_0^L and Y_0^L respectively. The main result is the next one.

Proposition 6.10
Both $\lim_{L\to\infty} w_L$ *and* $\lim_{n\to\infty} v_n$ *exist and coincide.*

Sketch of the proof
The result will follows from the inequality

$$\liminf_{n\to\infty} v_n \geq \limsup_{L\to\infty} w_L$$

and from the symmetry of the game. To each x in X_0^L one associates a strategy $\sigma(x)$ in G_n and in a dual way, any τ in G_n will be represented as some $y(\tau)$ in $\mathcal{G}_L$ in such a manner that, given $\varepsilon > 0$, $L \geq L(\varepsilon)$ and $n \geq N(\varepsilon, L, x)$ one obtains:

$$\overline{\gamma}_n(\sigma(x), \tau) \geq F_L(x, y(\tau)) - 0(\varepsilon).$$

Choose then $L \geq L(\varepsilon)$ such that w_L achieves the $\limsup_{L\to\infty} w_l$ up to ε, then use Ξ, ε-optimal with finite support on X_0^L to select x and play $\sigma(x)$. Explicitly the set of stages until n is divided into L large blocks and $x(\ell)$ defines the strategy played on block ℓ: uses $\alpha^{x(\ell)}$ i.i.d. except on $c(x(\ell))$ exceptional times uniformly distributed over this block. As for τ, pure strategy of Player 2, one uses a threshold r to define the representation as $y(\ell) = (\beta^{y(\ell)}, \{d_j^{y(\ell)}\})$ on each block.
The formal proof relies on two properties: First a good approximation of the revealing probability, in the spirit of Lemma 6.9, then an approximation of

the payoffs. The argument is the following. On the blocks where the probability of revelation is small the average payoff will be near one of the form $f^k(x(\ell), y(\ell))$. On the other hand, the number of blocks with non negligible revelation is uniformly bounded, before the payoff being essentially 0.

∎

The same kind of proof also shows that the above limit is $\lim_{\lambda\to 0} v_\lambda$. Remark that no explicit formula for this common limit is yet available.

6.4 Stochastic games with signals

This title covers an extension of the model of stochastic games presented in Chapter 5. It corresponds to the case where concerning the players'information, signals replace moves (see also 3.7.4 and 4.7.2), but where the state is still public knowledge at each stage.
Another extension dealing with games where the state is no longer publicly known is considered in the next Section 6.5.

6.4.1 Signalling structure

Consider a finite stochastic game with move sets I and J, state space Ω, payoff g and transition q as in Section 5.1. Signalling functions are mappings ℓ from $\Omega\times I\times J$ to $\Delta(A)$ (resp. m from $\Omega\times I\times J$ to $\Delta(B)$) where A and B are finite signals sets. At stage n, a signal a_n (resp. b_n) is chosen at random according to the distribution $\ell(\omega_n, i_n, j_n)$ (resp. $m(\omega_n, i_n, j_n)$) and announced to Player 1 (resp. Player 2) in addition to the new state ω_{n+1}. Assume that the signal to each player contains his move. Histories are now private and described at stage n by a point in $\Omega\times(A\times\Omega)^{n-1}$ for Player 1 and in $\Omega\times(B\times\Omega)^{n-1}$ for Player 2. Strategies are defined accordingly. (More generally one could consider a map Q from $\Omega\times I\times J$ to $\Delta(\Omega\times A\times B)$ to generate the triple (ω, a, b)).
Following the analysis in Section 5.2, v_n and v_λ are independent of the signalling structure. The interest is thus in the uniform approach.
Recall that already in the "Big Match" case (Section 5.3), the specification of the signalling structure is crucial: if Player 1 is not informed of the moves of Player 2, the max min is 0.
Most of the analysis presented below concerns absorbing games, hence in particular the signal can be assumed to be independent of the state.
The absorbing game is specified as follows. Given an entry (i, j), ω_{ij} is the probability of absorption, a_{ij} the absorbing payoff and b_{ij} the non absorbing payoff. Let r on $\Delta(I)\times\Delta(J)$ be defined by:

$$r(x,y) = \begin{cases} a^*(x,y) = \dfrac{\sum_{ij} \omega_{ij} x_i y_j a_{ij}}{\sum_{ij} \omega_{ij} x_i y_j} & \text{if } \omega(x,y) = \sum_{ij} \omega_{ij} x_i y_j > 0 \\ b(x,y) = \sum_{ij} x_i y_j b_{ij} & \text{otherwise} \end{cases}$$

so that $r(x,y)$ is the expected non absorbing payoff if (x,y) is non absorbing and otherwise corresponds to the normalized absorbing component of the expected payoff.

6.4.2 Absorbing games with no signals

In this subsection the signalling maps are $\ell(i,j) = i$ and $m(i,j) = j$ meaning that the players do not receive any outside information.

Proposition 6.11

$$\underline{v} = \sup_{x \in \Delta(I)} \inf_{y \in \Delta(J)} r(x,y)$$

Sketch of proof

For Player 1, choose an ε-optimal strategy for r and play it i.i.d..

As for Player 2, a first attempt would be to play at each stage a best reply to the expected strategy of Player 1 (conditionally to non absorption up to now). However this may not be sufficient if the corresponding payoff has an absorbing component and the total probability of absorption is less than one. One uses then the following property.

Lemma 6.12

$\forall \delta > 0$, $\exists \varepsilon_0$ and K such that for any x and $\varepsilon \le \varepsilon_0$ one has: if $r(x,j') \le \underline{v} + \delta$ implies $0 < \omega(x,j') \le \varepsilon$, then there exists some j with $\omega(x,j) \le K\varepsilon$ and $b(x,j) \le \underline{v} + 2\delta$.

In fact let J' be the set of j such that $g(x,j') \le \underline{v} + \delta$. Consider the normalization x_1 of x on $I \setminus I'$ with $I' = \{i{\in}I; \omega_{ij} > 0 \text{ for some } j{\in}J'\}$. If there is a non absorbing best reply to x_1 the assertion holds. Otherwise the absorbing probability is bounded by a constant times ε and one starts the procedure again.

This allows Player 2 to change his strategy, from j' to j, when the probability of future absorbtion is less than ε, and this without altering the underlying stochastic process by more than $K\varepsilon$. Player 2 uses thus first a best reply against the absorbing component and then switch to a best reply against the non absorbing one. ∎

6.4.3 Absorbing games with signalling structure

The following version of the "Big Match" (Section 5.3) will give a flavour of some of the new aspects due to the general signalling maps. One adds a dominated column (C is a large number) for Player 2 and the signalling mapping of Player 1 is given by the following:

payoffs =

1*	0*	C
0	1	C

signals =

?	?	?
a	b	a

It is easy to see that the min max is still 1/2 but now the max min is 0. In fact by playing $(0, \varepsilon, 1-\varepsilon)$ i.i.d. Player 2 generates a distribution $(1-\varepsilon, \varepsilon)$ on (a, b). He can play so until exhausting the probability of *Top* from Player 1 and then switch to $(1-\varepsilon, \varepsilon, 0)$ without being detected. During the first phase, the payoff is absorbing and equal to 0, and non absorbing and still 0 during the second one.

Before dealing with the general case it will be enlightening to consider the following class of games:

T	b_1	$\dots$	b_{J_1}	$\omega_{J_1+1}, a_{J_1+1}$	$\dots$	ω_{J_2}, a_{J_2}	$\omega_{J_2+1}, a_{J_2+1}$	$\dots$	ω_J, a_J
B	b_1	$\dots$	b_{J_1}	b_{J_1+1}	$\dots$	b_{J_2}	$\omega_{J_2+1}, a_{J_2+1}$	$\dots$	ω_J, a_J

The set of moves of Player 2 consists of 3 parts: $J = J_1 \cup J_2 \cup J_3$. FGiven $j \in J_1$ the payoff is non absorbing and equal to b_j. For $j \in J_2$ the payoff is absorbing with probability $\omega_j > 0$ and equal to a_j if Player 1 plays T and non absorbing and equal to b_j otherwise. Finally on J_3 the payoff is always absorbing with probability $\omega_j > 0$ and equal to a_j.

Define an equivalence relation on the mixed moves of Player 2 by: $y \sim y'$ if y and y' coïncide on J_3 and $\ell(B, y) = \ell(B, y')$.

Proposition 6.13
In the game above the minmax exists and is given by:

$$\underline{v} = \inf_{y \sim y'} \max\{r(T, y), r(B, y')\}$$

Sketch of proof
The fact that Player 2 can defend this amount is in the spirit of the above example. Given the stage behavior of Player 1, Player 2 has a reply y that guarantees $\underline{v}$. When the probability given σ and y to get absorption in the future is small, he will switch to y'.

The construction for Player 1 is much more intricate and relies on two main tools. One is an evaluation of the payoffs. Since the moves are not observable, Player 1 will make statistics on large blocks: based on the empirical frequency of signals he will deduce confidence lower bound ζ on the absorbing payoffs compatible with the strategy. This variable will then play the rôle of the statistics z in the proof in the standard signalling case (Proposition 5.7). ∎

The proof in the general case builds on the previous constructions and results. Given x and x' in X define as follows a game $C(x, x')$ of the form above. The B row and the signalling function are induced by the mixed move x in the original game. The T row differs from the B row on the moves $j \in J$ that are non absorbing for x but absorbing for x'. This defines the set J_2 and the corresponding payoff is the absorbing payoff facing x'. Denote by $\underline{v}(x, x')$ the minmax of this game, as defined by Proposition 6.13.

Proposition 6.14
Given an absorbing game with a signalling structure, the minmax exists and is given by:

$$\underline{v} = \sup_{x,x'} \underline{v}(x, x').$$

In particular it is independent of the information mapping of Player 2.

Sketch of the proof
The fact that Player 1 can guarantee this amount relies on the observation that he can generate the auxiliary game $C(x, x')$ by playing x when B is required and a mixture $(1-\varepsilon)x + \varepsilon x'$ when he is supposed to play T, where $\varepsilon > 0$ is small.
The proof for Player 2 is an extension of the idea of lemma 6.12. One shows that for any x of Player 1 such that the only best replies y of player 2 are absorbing with small probability, there is a decomposition $x = \alpha x(1) + (1-\alpha)x(2)$ and y' such that: y and y' are non aborbing and equivalent for $x(2)$ $(\ell(x(2), y) = \ell(x(2), y'))$ and y' is almost optimal facing $x(2)$. Player 2 plays then a best reply to the average strategy of Player 1, conditional to non absorption. If, at some point the best reply is still absorbing but with a small probability, he uses the previous decomposition to switch to y'. Here $x(1)$ plays the rôle of x and $x(2)$ the rôle of x' and the result follows.

∎

6.5 Stochastic games with incomplete information

6.5.1 Introduction

This last section is devoted to models described as stochastic games but where the state is not known by all the players. Most of the cases will belong to the framework of two-person zero-sum games with lack of information on one side and standard signalling: Player 1 knows the whole past history while Player 2 does not.

Some tools and results extend easily from Chapters 3 and 5. On the other hand there are also some fundamental differences with classical stochastic games or games with incomplete information on one side: non existence of the value, non-algebraicity of the minmax, maxmin or $\lim v_n$.

6.5.2 A first model

The game is specified by a state space Ω, a map q from $\Omega \times I \times J$ to probabilities on $\Omega \times M$ where M is a set of public signals. It is assumed that each signal includes the move of Player 2, hence m determines $j = j(m)$. p is a probability on Ω according to which the initial state ω_1 is chosen. It is then announced to Player 1, while Player 2 knows only p. At stage n, the state ω_n

and the moves i_n and j_n determine the payoff $g_n = g(\omega_n, i_n, j_n)$. $q(\omega_n, i_n, j_n)$ is the law of (ω_{n+1}, m_{n+1}) where ω_{n+1} is the new state and m_{n+1} the public signal to both players. Write as usual $G_n(p)$ (resp. $G_\lambda(p)$) for the n-stage (resp. λ-discounted) version of the game. Stage after stage, the strategy of Player 1 and the public signal determine a posterior distribution on Ω, hence a recursive formula for v_n and v_λ holds and properties of optimal strategies of Player 1 obtain, see Section 6.3. The natural state space is thus $\Delta(\Omega)$.

A specific class (level 1) corresponds to a finite family G^k, $k \in K$, of stochastic games on the same state space Ξ and with the same move spaces I and J. Hence $\Omega = K \times \Xi$. $\xi_1 \in \Xi$ is given and known by the players. k is chosen according to some initial probability p known also by both. Player 1 is informed of k but not Player 2. Then the game G^k is played with at each stage a new state in Ξ and a new public signal in M determined by $q^k(\xi, i, j)$ in $\Delta(\Xi \times M)$. We also assume that the signal contains the new state in Ξ and the move of Player 2. Note that since the transition depends on k the signal may be revealing on K even if Player 1 plays without using his information ($x^{k,\xi} = x^{k',\xi}$). The main difference with the previous case is that the beliefs at each stage are of the form $(\tilde{p}, \tilde{\xi})$ where $\tilde{p}$ is a martingale on K. The natural state space is $\Delta(K) \times \Xi$.

A further subclass (level 2) assume that the law of (ξ, m) is independent of k. Basically the game is then a (usual) stochastic game with state space Ξ but vector payoffs in $\mathbb{R}^K$. Player 1 knows the true component while Player 2 has only an initial probability on it.

6.5.3 Recursive formula

Given an initial probability p, a one stage strategy $\{x^\omega\}_{\omega \in \Omega}$, with $x^\omega \in \Delta(I)$, and a signal m, define the conditional probability on Ω given m by: $\tilde{p}^\omega(m) = Prob(\omega | m)$, with $Prob(\omega, m) = \sum_{\omega', i} p^{\omega'} x^{\omega'}(i) q(\omega, m | \omega', i, j(m))$ and $Prob(m) = \sum_\omega Prob(\omega, m)$. Then the usual recursive formula holds:

$$nv_n(p) = \max_{X^\Omega} \min_{J} \left\{ \sum\nolimits_{\omega, i} p^\omega x^\omega(i) g(\omega, i, j) + (n-1) E_{p,x,j}[v_{n-1}(\tilde{p}(m))] \right\}$$

and similarly for $v_\lambda(p)$. This implies that Player 1 has an optimal Markov strategy in $G_n(p)$ and an optimal Markov stationary strategy in $G_\lambda(p)$ where the state space is $\Delta(\Omega)$. Recall that, on the countrary, the recursive formula does not allow to construct recursively optimal strategies of Player 2 since it involves the computation of the posterior distribution, unknown to Player 2 who ignores x. Also results similar to those of Subsection 3.5.1. hold. First $v_n(p)$ and $v_\lambda(p)$ are concave on $\Delta(\Omega)$. Then the dual game is introduced. Defining $\pi \in \Pi = \Delta(\Omega \times I)$, where the marginal π_Ω on Ω plays the role of p and the conditional on I given ω the role of x^ω, one obtains:

$$\begin{aligned} nv_n^*(\zeta) = \max_{\pi \in \Pi} \min_{y \in Y} \{ & \sum\nolimits_{\omega, i, j} \pi(\omega, i) y(j) g(\omega, i, j) \\ & + (n-1) E_{\pi, y}[v_{n-1}(\tilde{\pi}_\Omega(m))] - n\langle \zeta, \pi_\Omega \rangle \}. \end{aligned} \tag{6.2}$$

where $\tilde{\pi}_\Omega(m)$ is the conditional distribution on Ω, given ω.
When specialized atlLevel 2 this describe an optimal strategy of Player 2 in the dual game which is Markov on $\Xi \times \mathbb{R}^K$. By Corollary 2.10 a similar property holds in the primal game.

6.5.4 Recursive games with incomplete information

Recursive game with absorbing payoffs are considered: the payoff is either 0 or absorbing. Denote by S the set of non absorbing states, by $A = \{-1, 1\}$ the set of absorbing states/payoffs and by p the initial probability on S according to which the state is chosen and announced to Player 1. At each stage n given the strategy of Player 1, Player 2 computes the posterior probability ρ_n on S, conditional to the fact that the payoff is not absorbing. Note that if only the state were absorbing, Player 2 should know which state is reached in order to play well afterwards. Here, in case an absorbing state is reached the future payoff is independent of the move, hence Player 2 can concentrate on the complementary event. If the stat eat stage n is absorbing ρ_n is the Dirac mass at that point.
The α-weighted Shapley operator (Section 5.2) is defined on $\Delta(S)$ by:

$$\Phi(\alpha, f)(p) = \mathtt{val}_{X^S \times Y} \{(1-\alpha)(\pi(p,x,y)E(f(\rho)) + (1 - \pi(p,x,y))E(a))\}$$

where $\pi(p,x,y)$ is the probability to remain in S and ρ the corresponding posterior probability. a stands for the absorbing payoff. Obviously $v_n(a) = v_\lambda(a) = v_\infty(a) = a$ on A and we consider their asymptotic behavior on $\Delta(S)$.

Theorem 6.15

$$\max\min = \lim_{n\to\infty} v_n = \lim_{\lambda\to 0} v_\lambda$$

Sketch of Proof
For any p, $v_\lambda(p)$ is well defined. Let w an accumulation point of the family $\{v_\lambda\}$, which is uniformly Lipschitz on $\Delta(S)$. As usual:

$$\Phi(0, w) = w$$

Let us show that Player 1 can guarantee w. As long as w is negative, playing optimal in the "projective game" corresponding to $\Phi(0, w)$ will guarantee w since the current payoff is $0 \geq w$. However the argument fails if $w(p) > 0$ and the idea is then to play optimal in a λ discounted game with $\|v_\lambda - w\|$ small. Given $\varepsilon > 0$, let λ such that $\|v_\lambda - w\| \leq \varepsilon^2$. Write $\mathbf{x}_\lambda(p)$ (resp. $\mathbf{x}(p)$) for an optimal strategy of Player 1 in $\Phi(\lambda, v_\lambda)(p)$ (resp. $\Phi(0, w)(p)$). Inductively a strategy σ of Player 1 and stopping times θ_ℓ are defined as follows:
Let $\theta_1 = \min\{m : w(\rho_m) > \varepsilon\}$ and play $\mathbf{x}(\rho_n)$ at each stage n until θ_1 (excluded). Let then $\theta_2 = \min\{m \geq \theta_1; v_\lambda(\rho_m) < 0\}$ and play $\mathbf{x}_\lambda(\rho_n)$ at each stage n from θ_1 until θ_2(excluded). More generally play $\mathbf{x}(\rho_n)$ from stage $\theta_{2\ell}$ to $\theta_{2\ell+1} = \min\{m : w(\rho_m) > \varepsilon\}$(excluded) and play $\mathbf{x}_\lambda(\rho_n)$ from $\theta_{2\ell+1}$ until

$\theta_{2\ell+2} = \min\{m; v_\lambda(\rho_m) < 0\}$(excluded).
Let u_n be equal to $w(\rho_n)$ at nodes where Player 1 is using $\mathbf{x}$ (i.e. playing optimally for $\Phi(0,w)$) hence for $\theta_{2\ell} \leq n < \theta_{2\ell+1}$ and be equal to $v_\lambda(\rho_n)$ otherwise. We call the first set of nodes "increasing" and the other set "decreasing".
A first property is that u_n is essentially a submartingale. This is clear if one starts at an increasing node and stays in this set since by the choice of σ:

$$E_{\sigma,\tau}(u_{n+1}|\mathcal{H}_n) = E_{\sigma,\tau}(w|\mathcal{H}_n) \geq \Phi(0,w) = w = u_n.$$

Similarly if the initial node is decreasing and one remains in this set, one obtains using the fact that $v_\lambda(p_n) \geq 0$ (by the choice of the stopping time):

$$E_{\sigma,\tau}((1-\lambda)u_{n+1}|\mathcal{H}_n) = E_{\sigma,\tau}((1-\lambda)v_\lambda|\mathcal{H}_n) \geq \Phi(\lambda, v_\lambda) = v_\lambda = u_n \geq 0$$

so that

$$E_{\sigma,\tau}(u_{n+1}|\mathcal{H}_n) \geq u_n.$$

Now if one of the new node changes from decreasing to increasing or reciprocally, the error is at most ε^2 hence in all cases:

$$E_{\sigma,\tau}(u_{n+1}|\mathcal{H}_n) \geq u_n - \varepsilon^2 P(n+1 \in \Theta|\mathcal{H}_n)$$

where Θ is the random set of all values of stopping times $\{\theta_\ell\}$.
A second property is a bound on the error term using the fact that the stopping times count the upcrossing of the band $[0,\varepsilon]$ by the sequence u_n. If η^N denotes the number of stopping times θ_ℓ before stage N and $\eta = \lim \eta^N$ one has:

$$E(\eta) \leq \frac{2}{\varepsilon - \varepsilon^2}$$

and one uses:

$$\sum\nolimits_n P(n+1 \in \Theta) = \sum\nolimits_\ell \sum\nolimits_{n+1} P(\theta_\ell = n+1) \leq E(\eta) + 1$$

to get finally

$$E(u_n) \geq u_1 - 5\varepsilon.$$

The last point is to compare u_n to the current payoff in the game. Until absorbtion the current payoff is 0 hence near w (or v_λ) as long as $w \leq \varepsilon$. Define $\mathcal{A}_n$ to be the set of non absorbing nodes with $w(\rho_n) > \varepsilon$. One obtains:

$$E(g_n) \geq u_1 - 7\varepsilon - 2P(\mathcal{A}_n).$$

Denoting by ξ the absorbing time the crucial property is that $\forall \varepsilon > 0, \lambda > 0, \exists N$ such that:

$$P(\xi \leq n + N|\mathcal{A}_n) \geq \varepsilon/2.$$

This result follows from the fact that given a node in $\mathcal{A}_n$, Player 1 is using $\mathbf{x}_\lambda$ as long as $v_\lambda(\rho_m) \geq 0$. Now before absorption $E((1-\lambda)v_\lambda) \geq v_\lambda$. Since v_λ is

bounded, positive and increasing geometrically there is a positive probability of absorption in finite time.
One then deduces that $\sum_n P(\mathcal{A}_n)$ is uniformly bounded, hence:

$$E(\overline{g}_n) \geq w - 8\varepsilon$$

for n large enough. Since the strategy of Player 1 is independent of the length of the game, this implies that Player 1 can guarantee w.
Given any strategy σ of Player 1, Player 2 can compute the posterior distribution ρ_n as well and use the "dual" of the previous strategy. The same bound (independent of σ) thus implies that $\max\min = w$ and moreover $\lim v_n = \lim v_\lambda = w$.

∎

An example is given in Rosenberg and Vieille (2000) where they show that max min and min max may differ.
On the other hand the previous proof shows that the crucial point is the knowledge of the belief parameters, which is the case in the asymptotic analysis. Hence one obtains:

Proposition 6.16
Consider a recursive game with absorbing payoffs and lack of information on both sides. Then both $\lim_{n\to\infty} v_n$ *and* $= \lim_{\lambda\to 0} v_\lambda$ *exist and coincide.*

6.5.5 Absorbing games with i.i.

This corresponds to a subclass of level 2: absorbing games with vector payoffs. A recent and important result by Rosenberg (1999) is worth mentioning.

Theorem 6.17
For absorbing games with vector payoffs and incomplete information on one side, both $\lim v_n$ *and* $\lim v_\lambda$ *exist and coincide.*

The proof is very involved and uses the operator approach (Appendix C.3) to obtain variational properties satisfied by any accumulation point of the family $\{v_\lambda\}$ and then deduce uniqueness.

The remaining of this section is a collection of partial results introducing new tools and ideas that may be useful in more general cases. The games under consideration have a structure similar to the Big Match of Blackwell and Ferguson (1968), see Section 5.3, namely these are absorbing games where one of the player controls the absorption. However there is some incomplete information on the state, hence the name for this class.

6.5.6 "Big Match" with i.i.: type I

Consider a family of games of the following form:

$$G^k = \begin{pmatrix} a_1^{k*} & a_2^{k*} & \dots & a_m^{k*} \\ b_1^k & b_2^k & \dots & b_m^k \end{pmatrix}$$

where the first line is absorbing. k belongs to a finite set K and is selected according to p in $\Delta(K)$. Player 1 knows k while Player 2 knows only p.

A. Asymptotic analysis

The use of the recursive formula allows to deduce properties of optimal strategies. In particular in our case the value of the game is the same if both players are restricted to strategies independent of the past: first the information transmitted to Player 2 is independent of Player 2's own moves, hence one can ignore them; second, there is only one past history of moves of Player 1 to take into consideration, namely Bottom up to the current stage (excluded). This suggests to construct an asymptotic game $\wp(p)$ played between time 0 and 1 and described as follows (compare with Subsection 5.3.2):

ρ^k is the law of the stopping time θ corresponding to the first stage where Player 1 plays Top, if k is announced: $\rho^k(t) = Prob_{\sigma^k}(\theta \le t)$.

f is a map from $[0,1]$ to $\Delta(J)$, $f(t)$ being the mixed strategy used by Player 2 at time t.

The payoff is given by $L(\{\rho\}, f) = \sum_{k\in K} p^k L^k(\rho^k, f)$ where L^k is the payoff in game k, expressed as the integral between 0 and 1 of the "payoff at time t":

$$L^k(\rho^k, f) = \int_0^1 L_t^k(\rho^k, f)dt$$

with, letting $A^k f = \sum_{j\in J} a_j^k f_j$ and similarly for $B^k f$, the following expression for L_t^k:

$$L_t^k(\rho^k, f) = \int_0^t A^k f(s)\rho^k(ds) + \left(1 - \rho^k(t)\right) B^k f(t).$$

The first term corresponds to the absorbing component of the payoff and the second one to the non absorbing one.

Theorem 6.18

1) The game $\wp(p)$ has a value $w(p)$ on $\Delta(K)$.

2) $\lim_{n\to\infty} v_n = \lim_{\lambda\to 0} v_\lambda = w$.

Proof

The existence of a value follows from Sion's minmax theorem.

Consider now ε−optimal strategies ($\rho = \{\rho^k\}, f$) in $\mathcal{G}$. They induce natural discretisations $(\rho(n), f(n))$ or $(\rho(\lambda), f(\lambda))$ in G_n or G_λ corresponding to piece wise constant approximations on the intervals of the form $[\frac{m}{n}, \frac{m+1}{n}]$ or

$[\sum_{t=1}^{m-1}\lambda\,(1-\lambda)^{t-1}, \sum_{t=1}^{m}\lambda(1-\lambda)^{t-1}]$. Continuity of the payoffs imply that $\rho(n)$ will guarantee w up to some constant times $(1/n+\varepsilon)$ in G_n and a dual result holds for Player 2 using $f(n)$. A similar property is obtained for G_λ.

∎

B. Maxmin

The construction relies on properties of the previous auxiliary game $\wp$ and the result is the following:

Theorem 6.19

$$maxmin = w$$

Proof

We first prove that Player 2 can defend w.

Let f be an ε−optimal strategy of Player 2 in $\wp$. Player 2 will mimick f in G_∞ in order to generate through the strategy σ of player 1 a family of distributions $\{\mu^k\}$ such that by playing "up to level t" the payoff will be near $L_t(\{\mu\}, f)$. Since by the choice of f, $L(\{\mu\}, f) = \int_0^1 L_t(\{\mu\}, f)$ is less than $\omega + \varepsilon$, there exists some t^* with $L_{t^*}(\{\mu\}, f) \leq w + \varepsilon$. This will define the strategy τ of Player 2 as: follow f up to level t^*.

Formally we consider a discrete valued ε−approximation $\tilde{f}$ of f, $\tilde{f}$ being equal to f_i on $[t_i, t_{i+1}[$, with $i \in I$, finite. Given σ, the positive measures μ^k are defined inductively as follows:

$\overline{\mu}^k(t_1) = Prob_{\sigma^k,\tau_1}(\theta < +\infty)$ where τ_1 is f_1 i.i.d..

Let N_1 be such that the above probabilities are almost achieved by that stage for all k, this defines $\mu^k(t_1) = Prob_{\sigma,\tau_1}(\theta \leq N_1)$.

$\overline{\mu}^k(t_2) = Prob_{\sigma^k,\tau_2}(\theta < +\infty)$, where τ_2 is τ_1 up to stage N_1 and then f_2 i.i.d..

One introduces N_2, τ_2 as above and so on:

$\overline{\mu}^k(t_i) = Prob_{\sigma^k,\tau_i}(\theta < +\infty)$ where τ_i is τ_{i-1} up to N_{i-1} and then f_1 i.i.d..

It is then clear that the payoff induced in G_n, for n large enough, by σ and τ_i will be of the form:

$$\sum\nolimits_k p^k \left\{\sum\nolimits_{j\leq i}(\mu^k(t_j) - \mu^k(t_{j-1}))A^k f_j + (1-\mu^k(t_i))B^k f_i\right\}$$

hence near $L_{t_i}(\{\mu\}, \tilde{f})$. Since $\int_0^1 L_t(\mu, \tilde{f})dt$ is at most w (up to some approximation), there exists an index i^* with $L_{t_{i^*}}(\{\mu\}, \tilde{f})$ below $w + O(\varepsilon)$. Finally the strategy τ_{i^*} defends w.

The proof that Player 1 can guarantee w is more intricate.

We first show the existence of a couple of optimal strategies $(\rho = \{\rho^k\}, f)$ in $\wp$ that are essentially equalizing, namely such that $L_t(\{\rho\}, f)$ is near w for all t. In fact consider $\{\rho\}$, optimal for Player 1 in $\wp$ and the game $\wp_0$ where Player 1 chooses t, Player 2 chooses f and the payoff is $L_t(\{\rho\}, f)$.

Proposition 6.20

The game $\wp_0$ has a value, which is the same as for $\wp$, w.

Proof
The existence of a value, w_0, follows again from Sion's theorem. Since Player 1 can choose the uniform distribution on $[0,1]$ which induces $L(\{\rho\}, f)$, the value w_0 is at least w. If $w_0 > w$, an optimal strategy of Player 1, hence a cumulative distribution function on $[0,1]$, α, could be used to make a "change of time" and to induce in $\wp$ through the image of ρ by α a payoff always at least w_0. ∎

The idea of the proof is now to follow the "path defined by f and ρ". Basically, given k, Player 1 will choose t according to the distribution ρ^k and play the strategy δ_t where δ_t is defined inductively as follows: consider the non absorbing payoff at time t

$$\sum_k p^k(1-\mu^k(t))B^k f(t) = b_t(f)$$

Player 1 uses then a "Big Match" strategy blocking whenever the non absorbing payoff evaluated trough $b_t(.)$ is less than $b_t(f)$. The equalizing property of f then implies that the absorbing payoff will be at least the one corresponding to f. It follows that the total payoff is minorized by an expectation of terms of the form $L_t(\{\rho\}, f)$, hence the result. ∎

C. Minmax
This last part is much simpler.

Theorem 6.21
minmax $= v_1$, *value of the one stage game.*

Proof
It is clear that by playing i.i.d. an optimal strategy y in the one stage game Player 2 will induce an expected payoff at any stage n of the form:

$$g_1(p;\alpha,y) = \sum_k p^k(\alpha^k A^k y + (1-\alpha^k)B^k y)$$

where $\alpha^k = Prob_{\sigma^k,\tau}(\theta \le n)$, hence less than v_1.
To prove that Player 1 can defend v_1, let $\alpha = \{\alpha^k\}$ be an optimal strategy for him in $G_1(p)$. Knowing τ, Player 1 evaluates the non absorbing component of the payoff at stage n given α namely:

$$c_n = \sum_k p^k(1-\alpha^k)B^k \overline{y}_n$$

where $\overline{y}_n = E(\tau(h_n)|\theta \ge n)$ is the expected mixed move of Player 2 at stage n, conditional to Player 1 playing Bottom up to that stage. Letting c_N achieve up to ε the supremum of c_n, Player 1 plays Bottom up to stage N excluded, then once α at stage N and always Bottom there after. For n larger than N the expected payoff will be of the form:

$$\sum_k p^k(\alpha^k A^k \overline{y}_N) + c_n$$

hence greater than $g_1(p;\alpha,\overline{y}_N) - \varepsilon$, which gives the result.

■

Example

Consider the following game with $p = p^1 = Prob(G^1)$:

$$G^1 = \begin{pmatrix} 1^* & 0^* \\ 0 & 0 \end{pmatrix} \qquad G^2 = \begin{pmatrix} 0^* & 0^* \\ 0 & 1 \end{pmatrix}$$

Then one obtains that $w(p)$ can be written as:

$$w(p) = \inf_{f:[0,1]\to[0,1]} \sup_{t\in[0,1]} \{p(1-t)f(t) + (1-p)\int_0^1 (1-f(s))ds\}$$

from which it follows that:

$$\underline{v}(p) = \lim_{n\to\infty} v_n(p) = \lim_{\lambda\to 0} v_\lambda(p) = w(p) = (1-p)(1-exp(-\frac{p}{1-p}))$$

$$\overline{v}(p) = v_1(p) = \min(p, 1-p)$$

In particular, the uniform value does not exist, the asymptotic value and the maxmin are transcendental functions: at $p = \frac{1}{2}$ one obtains $\underline{v}(\frac{1}{2}) = \frac{1}{2}(1 - \frac{1}{e})$ while all the data are rational numbers.

D. Extensions

We study here the extension to absorbing states rather than absorbing payoffs. The games are of the form:

$$\begin{pmatrix} A_1^{k*} & A_2^{k*} & \dots & A_m^{k*} & \dots \\ b_1^k & b_2^k & \dots & b_m^k & \dots \end{pmatrix}$$

where $A_1 = \{A_1^k\}$, ... $A_m = \{A_m^k\}$, ... are games with incomplete information corresponding to absorbing states. It follows that when Player 1 plays Top, the state is absorbing but not the payoff and the strategic behavior there after (hence also the payoff) will be a function of the past. Let $v_m(p)$ be the uniform value of the game A_m with initial distribution p (Section 3.6). The recursive formula implies that the absorbing payoff is approximately $\sum_m v_m(p^T)y_m$ (where p^T is the conditional distribution given Top and y_m the probability of playing column m) if the number of stages that remains to be played is large enough.

Consider now the continuous time game $\wp$. Given a profile $\rho = \{\rho^k\}$, denote by $p^T(t)$ (resp. $p^B(t)$) the conditional probability on K given $\theta = t$ (resp. $\theta > t$). The payoff is defined as:

$$M(\{\rho\}, f) = \int_0^1 M_t(\{\rho\}, f)dt$$

where the payoff at time t is given by:

$$M_t(\{\rho\},f) = \int_0^t (\sum\nolimits_m v_m(p^T(s)) f_m(s)) d\overline{\rho}(s) + (1-\overline{\rho}(t))(\sum\nolimits_k p_k^B(t) b^k f(t))$$

$\overline{\rho}(t) = \sum_k p^k \rho^k(t)$ being the average probability of the event $\{\theta \leq t\}$.
M is still a concave function of ρ (due to the concavity of each v_m) and Sion's theorem still applies. The analogous of Theorem 6.19 then holds. One shows that Player 1 can obtain w in long games, and using the minmax theorem that he cannot get better. Similarly the analysis of the maxmin follows the same lines.
Concerning the maxmin one is led to introduce a family of games as follows: for each game A_m consider the set Ξ_m of vector payoffs (in $\mathbb{R}^K$) that Player 2 can approach, see Section 3.6.2, namely such that:

$$\langle p, \xi_m \rangle \geq v_m(p) \quad \forall p \in \Delta(K).$$

Given a profile $\xi = \{\xi_m\}$ of vectors in $\prod_m \Xi_m$ we consider the game $\mathcal{A}(\xi, p)$, where each component is given by:

$$\mathcal{A}^k(\xi) = \begin{pmatrix} \xi_1^{k*} & \xi_2^{k*} & \ldots & \xi_m^{k*} & \ldots \\ b_1^k & b_2^k & \ldots & b_m^k & \ldots \end{pmatrix}$$

By construction for each such ξ, Player 2 can guarantee (in the original game) the minmax of $\mathcal{A}(\xi, p)$ which is the value of the one shot version, say $\nu_1(\xi, p)$. One then has:

$$minmax = \min_{\xi \in \Xi} \nu_1(\xi, p).$$

In fact by playing optimally for the minmax in $\mathcal{A}(\xi, p)$, Player 1 is anticipating the behavior of Player 2, after absorption (namely approach ξ_m if absorption occured when playing m). The best Player 2 could do then would be to choose a supporting hyperplane to v_m at the current posterior p^T. This defines a correspondance C from $\prod_m \Xi_m$ to itself. One shows that C is u.s.c. with convex values hence has a fixed point ξ^*. Playing optimally for the minmax against τ in $\mathcal{A}(\xi^*, p)$ will then guarantee an absorbing payoff above ξ^* hence a total payoff above $\nu_1(\xi^*, p)$.

6.5.7 "Big Match" with i.i.: type II

We consider here games of the form:

$$G^k = \begin{pmatrix} a_1^* & b_1 \\ a_2^* & b_2 \\ \ldots & \ldots \\ a_m^* & b_m \\ \ldots & \ldots \end{pmatrix}$$

where the first column is absorbing: Player 2 controls the transition. As usual the game G^k is chosen with probability p^k and announced to Player 1.

A. Asymptotic analysis and maxmin

The analysis is roughly similar in both cases and based on the tools developed in Chapters 3 and 4 for incomplete information games. Let u be the value of the non-revealing game (where Player 1 is not transmitting any information on k). A crucial property is that this value does not depend upon the length of the game and then one shows immediately that Player 1 can guarantee $\mathtt{Cav}u\ (p)$, where $\mathtt{Cav}$ denotes the concavification operator on the simplex $\Delta(K)$. Since Player 2 has a "non absorbing" move he can (in the compact case or for the maxmin), knowing σ, observe the variation of the martingale of posteriors probabilities on K. Except for a vanishing fraction of stages this variation is small, hence Player 1 is almost playing non-revealing so that a best reply of Player 2 gives a payoff near u at the current posterior. The result follows by averaging in time and taking expectation, using Jensen's inequality. We thus obtain:

Theorem 6.22
$\max\min = \lim_{n\to\infty} v_n = \lim_{\lambda\to 0} v_\lambda$

B. Minmax

The analysis in this case requires quite specific tools and is related to the problem of approachability in stochastic games with vector payoffs. We will simply give hints on two examples.

Example 1

$$G^1 = \begin{pmatrix} 1^* & 0 \\ 0^* & 0 \end{pmatrix} \quad G^2 = \begin{pmatrix} 0^* & 0 \\ 0^* & 1 \end{pmatrix}$$

One easily has, with $p = Prob(k=1)$, that: $u(p) = p(1-p)$, hence:

$$\underline{v}(p) = \lim_{n\to\infty} v_n(p) = \lim_{\lambda\to 0} v_\lambda(p) = \mathtt{Cav}\ u(p) = p(1-p).$$

However

$$\overline{v}(p) = p(1 - exp(1 - \frac{(1-p)}{p}))$$

and is obtained as follows. Denote by $\beta_t, 0 \le t \le 1$, a "Big Match" type strategy of Player 2, see Section 5.3, absorbing when the frequency of Bottom exceeds t, namely optimal in the game:

$$\begin{pmatrix} t^* & -t \\ -(1-t)^* & (1-t) \end{pmatrix}$$

Player 2 chooses t according to some distribution ρ and then play β_t. A best reply of Player 1 is then to start by playing Top and to decreasing slowly his frequency of Top, in order to get an absorbing payoff as high as possible. This leads to the following quantity that Player 2 can guarantee:

$$\tilde{v}(p) = \inf_{\rho} \sup_{t} \{p \int_0^1 (1-s)\rho(ds) + (1-p)t(1-\rho([0,t])\}$$

To prove that Player 1 can defend $\tilde{v}$ let him construct such a measure ρ starting from the strategy τ of Player 2. A discretisation will be obtained by playing Bottom with frequency $\frac{\ell}{N}, \ell = 0, ..., N$, for N large. $\tilde{R}(0)$ is thus the probability of absorbtion given "always Top". It is almost achieved at stage N_0, this defines the quantity $R(0)$. Inductively $\tilde{R}(\ell)$ is the probability of absorbtion given the previous strategy until stage $N_{\ell-1}$ and then $(1-\frac{\ell}{N}, \frac{\ell}{N})$ i.i.d. By choosing ℓ and using the associated strategy Player 1 can thus achieve $\tilde{v}$.

Example 2

$$G^1 = \begin{pmatrix} 1^* & 0 \\ 0^* & 1 \end{pmatrix} \quad G^2 = \begin{pmatrix} 0^* & 3/4 \\ 1^* & 0 \end{pmatrix}$$

Given a point C in $\mathbb{R}^2$, to say that Player 2 can approach C means that he can approach the negative orthant with origin C, see Appendix B, namely: for any ε there exists τ and N such that for any σ: $\overline{\gamma}_n^k(\sigma, \tau) \leq C^k + \varepsilon$ for $n \geq N$.

Clearly Player 2 can approach $X = (1, 3/7)$ by playing optimally in G^2.

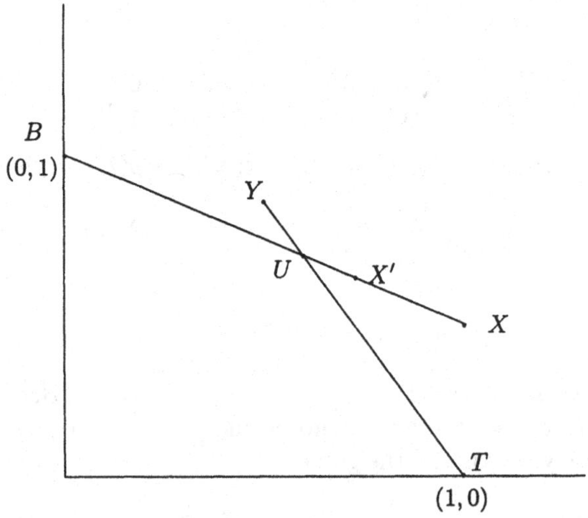

He can also approach $Y = (1/2, 3/4)$ by playing a sophisticated optimal strategy in G^1: start like an optimal strategy in G^1 but control both the

absorbtion probability (q) and the expected absorbing payoff (a) to satisfy $qa + (1 - q) \geq 1/2$: as soon as the opposite equality holds Player 2 can play anything in G^1 and get a payoff less than 1/2, in particular by playing optimally in G^2. This allows him to approach Y.
Let $T = (1,0)$ and $B = (0,1)$. We will show that Player 2 can also approach U which is the intersection of the segments $[BX]$ and $[TY]$. Note that $U = 1/13T + 12/13Y = 6/13B + 7/13X$. Player 2 plays then Left with probability 1/13. If Top is played the absorbing event is $(q, a) = (1/13, T)$, hence it remains to approach Y. Otherwise the absorbing event is $(q, a) = (1/13, B)$ hence it remains to approach X' with $U = 1/13B + 12/13X'$. Choose now a point U' on TY Pareto dominated by X' and start again.
An example of such procedure is given by:

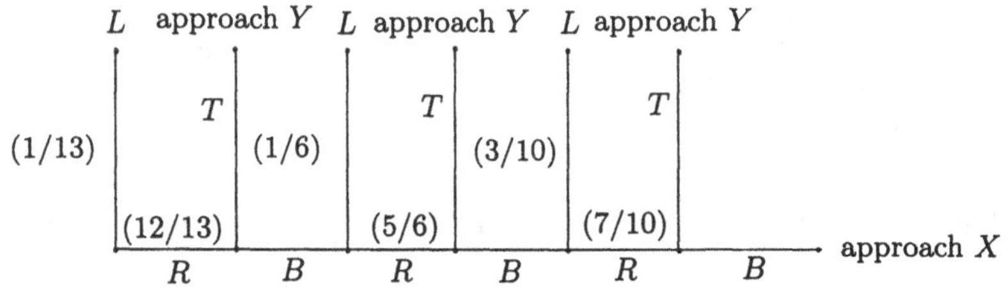

As for Player 1, by playing Top until exhausting the probability of absorbtion and then eventually optimal in G^1 he forces a vector payoff minorized by a point Z of the form : $\alpha T + (1 - \alpha)Y$, hence on $[TY]$. Similarly by playing Bottom and then eventually optimal in G^2, Player 1 can "defend" the payoffs above the segment $[XB]$.
Finally it is easy to see that the set of points that Player 2 can appraoch is convex and that similarly Player 1 can defend any convex combination of halfspaces that he can defend.
It follows that the "approachable set" is the set $\mathcal{C}$ of points C with $C^k \geq Z^k$ for some Z in the convex hull of (X, Y, U). Finally the maxmin $\overline{v}$ is simply the support function of $\mathcal{C}$:

$$\overline{v}(p) = \min_{C \in \mathcal{C}} \langle C, p \rangle$$

6.5.8 Comments

In all the cases studied up to now where Player 1 is more informed than Player 2, the maxmin is equal to the asymptotic value ($\lim v_n$ and $\lim v_\lambda$) and it is conjectured that this is a general property for this class. More intuition for this to hold can be obtained using the general recursive approach (see MSZ).

As mentioned in 4.7.2 the analysis of games with lack of information on both sides and state dependent signalling is quite preliminary. In this case even by playing whithout using his own information a player may reveal it. However two classes can be considered: in the first one the support of the signalling matrices is independent of the state and one can define non-revealing moves. The other one corresponds to "revealing signals". Absorbing games with incomplete information were introduced as auxiliary tools to study those games. An example is given by the following case, Sorin (1985b). The state is (k,ℓ). Player 1 knows k and player 2 knows ℓ. Each game $A^{k\ell}$ is 2×2 and the signalling matrices are as follows:

$$H^{11} = \begin{pmatrix} T & L \\ P & Q \end{pmatrix} \quad H^{12} = \begin{pmatrix} T & R \\ P & Q \end{pmatrix}$$

$$H^{21} = \begin{pmatrix} B & L \\ P & Q \end{pmatrix} \quad H^{22} = \begin{pmatrix} B & R \\ P & Q \end{pmatrix}$$

As soon as player 1 plays Top some game is revealed and one can assume the state absorbing. Then the analysis of the maxmin and the minmax uses tools in the spirit of 5.6 and 5.7.

An example analyzed by Ferguson, Shapley and Weber (1970) deals with two games G^1 and G^2 as follows:

$$G^1 = (1 \quad 0) \quad G^2 = \begin{pmatrix} 0 & 1 \\ 0 & 0 \end{pmatrix}$$

The transition from G^1 to G^2 is $1 > \pi > 0$ whatever being the moves and Player 1 controls the transition from G^2: stay if *Top* is played and move to G^1 if *Bottom* occured. Player 1 has complete information while Player 2 is basically only told when a transition from G^2 to G^1 occurs. The value is obtained through an auxiliary game where Player 2 is using strategies that are function of m, the number of stages since his last message.

Recall finally that Exercise 5.9.9 describes a one person stochastic game where the state is unknown.

6.6 Notes

The model of Section 2 was introduced by Kohlberg and Zamir (1974) who proved the result in the deterministic framework. The extension to the random case (Section 2.2) is due to Forges (1982). The alternative approach described in Section 2.3 follows Neyman and Sorin (1997, 1998) and the last extensions are in Geitner (1999).
The model of Section 3 is due to Mertens and Zamir (1976a). The analysis in Sections 3.2 and 3.3 follows Mertens and Zamir (1976a) and Waternaux (1983a, 1983b). Section 3.4 comes from Sorin (1989).

Most of Section 4 is due to Coulomb (1992, 1999, 2001).
The results of Section 5.4. are due to Rosenberg and Vieille (2000). They also give an example of a recursive game with incomplete information on one side without value. The content of Sections 5.6 and 5.7 is issued from Sorin (1984a) and (1985a) respectively.

A: Minmax theorems and duality

A.1 Definitions

A **two-person zero-sum game** is a triple $(g; X, Y)$ where g is a real function defined on a product space $X \times Y$. X (resp. Y) is the **strategy set** of the maximizer, Player 1 (resp. of the minimizer, Player 2) and g is the **payoff function**. One introduces:

$$\underline{v} = \sup_{x \in X} \inf_{y \in Y} g(x, y) \text{ and } \overline{v} = \inf_{y \in Y} \sup_{x \in X} g(x, y).$$

Note that the inequality $\underline{v} \leq \overline{v}$ always holds.
A **minmax theorem** gives conditions for the game to have a **value**, namely for the equality $\underline{v} = \overline{v} = \mathtt{val}_{X \times Y} g$ to be satisfied.
A strategy x of Player 1 is **ε-optimal** if:

$$g(x, y) \geq \underline{v} - \varepsilon, \quad \forall y \in Y.$$

Similarly a strategy y of Player 2 is **ε-optimal** if:

$$g(x, y) \leq \overline{v} + \varepsilon, \quad \forall x \in X.$$

It always exists for $\varepsilon > 0$. If it exists for $\varepsilon = 0$, it is called **optimal.**

Under suitable measurability requirements one defines a **mixed extension** of (g, X, Y) as a game $(\gamma, \Sigma, \mathcal{T})$ where Σ and $\mathcal{T}$ are convex subsets of the set of probabilities on X and Y, and the payoff γ is the bilinear extension:

$$\gamma(\sigma, \tau) = \int_{X \times Y} g(x, y) \sigma(dx) \otimes \tau(dy).$$

A.2 Elementary approach

We start with a basic result known as the "Alternative Theorem":

Theorem A.1
Let A be a $m \times n$ matrix and b a $m \times 1$ vector, both with real coefficients. Let:

$$E_1 = \{x \in \mathbb{R}^n ; Ax \geq b\}$$

$$E_2 = \{u \in \mathbb{R}^m : u \geq 0 \ , uA = 0, \langle u, b\rangle > 0\}.$$

Then one and only one of the two sets is non empty.

Proof

If both E_1 and E_2 are nonempty one obtains:

$$0 \ = \langle uA, x\rangle = \langle u, Ax\rangle \geq \langle u, b\rangle > 0$$

a contradiction.

Let us now show that if E_1 is empty, E_2 is not. The proof is by induction on the number ℓ of variables actually present in the constraints.

For $\ell = 0$, E_1 reduces to $\{0 \geq b\}$, hence one component satisfies $b_i > 0$ and one can choose u to be the corresponding unit vector.

Assuming the result true for ℓ variables, take $n = \ell + 1$ and denote by I_0, (resp. I_+, I_-) the set of lines of A where the coefficient of $x_{\ell+1}$ is 0 (resp. $> 0, < 0$). Each line $\alpha \in I_+$ is associated to a constraint of the form:

$$x_{\ell+1} \geq c_\alpha(x_1, ..., x_\ell)$$

and similarly for $\beta \in I_-$

$$x_{\ell+1} \leq c_\beta(x_1, ..., x_\ell).$$

It is clear that $E_1 = \emptyset$ if and only if the following set is empty:

$$E_1' = \{x \in \mathbb{R}^\ell : A_{I_0} x \geq b_{I_0}, c_\beta(x) \geq c_\alpha(x), \forall(\alpha, \beta) \in I_+ \times I_-\}.$$

By induction hypothesis, there exists a vector $v = (v_{I_0}, \{v_{\alpha\beta}\}) \geq 0$ with:

$$\sum\nolimits_{i \in I_0} v_i a_{ij} + \sum\nolimits_{(\alpha,\beta)\in I_+ \times I_-} v_{\alpha\beta} \left(\frac{a_{\alpha j}}{a_{\alpha \ell+1}} - \frac{a_{\beta j}}{a_{\beta \ell+1}} \right) = 0 \quad j = 1, ..., \ell$$

$$\sum\nolimits_{i \in I_0} v_i b_i + \sum\nolimits_{(\alpha,\beta)\in I_+ \times I_-} v_{\alpha\beta} \left(\frac{b_\alpha}{a_{\alpha \ell+1}} - \frac{b_\beta}{a_{\beta \ell+1}} \right) \ > 0.$$

Let us define u by :

$$u_i = v_i, \quad i \in I_0; \quad u_\alpha = \sum\nolimits_\beta \frac{v_{\alpha\beta}}{a_{\alpha\ell+1}}, \quad \alpha \in I_+; \quad u_\beta = \sum\nolimits_\alpha \frac{v_{\alpha\beta}}{a_{\beta\ell+1}}, \quad \beta \in I_-.$$

The previous inequalities imply that u belongs to E_2. ∎

The above proof by elimination (due to Fourier) shows that if E_i is non empty, one can find explicitly a solution that belongs to the field ($\mathbb{R}$ or $\mathbb{Q}$) of coefficients of the problem. This property extends to any ordered field (Weyl, 1950).

Corollary A.2

$$\{Ax \geq 0 \Rightarrow \langle f,x\rangle \geq 0,\ \forall x \in \mathbb{R}^n\} \Longleftrightarrow \{\exists u \in \mathbb{R}^m, u \geq 0, uA = f\}.$$

Proof
The implication from right to left is clear.
If the right hand side set is empty, one obtains by the previous Theorem A.1 the existence of $(p,q,r) \in \mathbb{R}^m_+ \times \mathbb{R}^n_+ \times \mathbb{R}^n_+$ with

$$p + A(q-r) = 0, \quad \langle f, q-r\rangle > 0$$

hence the choice of $x = r - q$ contradicts the property on the left. ∎

The next Corollary is known as Farkas's Lemma.

Corollary A.3
Assume $\{x \in \mathbb{R}^n; Ax \geq b\} \neq \emptyset$.
Then:

i) $$Ax \geq b \Rightarrow \langle f,x\rangle \geq c, \quad \forall x \in \mathbb{R}^n$$

and

ii) $$\exists u \in \mathbb{R}^m, u \geq 0, uA = f, \langle u,b\rangle \geq c$$

are equivalent.

Proof
As above *ii)* implies *i)*.
Assume that the set described by *ii)* is empty. By Theorem A.1 there exist $(p,q,r,t) \in \mathbb{R}^m_+ \times \mathbb{R}^n_+ \times \mathbb{R}^n_+ \times \mathbb{R}$ with

$$p + A(q-r) + tb = 0, \quad \langle f, q-r\rangle + tc > 0.$$

Either $t > 0$, and the choice of $x = \frac{r-q}{t}$ contradicts *i)*, or $t = 0$. Choose then x' with $Ax' \geq b$. For $a > 0$ large enough, $x = x' + a(r-q)$ satifies:

$$Ax = Ax' + aA(r-q) \geq b, \ \langle f,x\rangle = \langle f,x'\rangle + a\langle f, r-q\rangle < 0$$

a contradiction to *i)*. ∎

Definition
Consider the following linear programs:

$$(\mathcal{P}_1) \quad \begin{array}{ll} \min & \langle c,x\rangle \\ & Ax \geq b \\ & x \geq 0 \end{array} \qquad\qquad (\mathcal{P}_2) \quad \begin{array}{ll} \max & \langle u,b\rangle \\ & uA \leq c \\ & u \geq 0 \end{array}$$

$(\mathcal{P}_2)$ is defined as the *dual* of $(\mathcal{P}_1)$. Note that the dual of $(\mathcal{P}_2)$ is $(\mathcal{P}_1)$.

Theorem A.4
If both programs $(\mathcal{P}_1)$ and $(\mathcal{P}_2)$ are feasible they have the same value.

Proof
Note first that for any feasible couple (x, u) one has:

$$\langle u, b\rangle \leq uAx \leq \langle c, x\rangle$$

hence both programs are bounded.
Let us now prove that $(\mathcal{P}_1)$ has a solution. If not, denoting by α its value, the set $E_1(\alpha) = \{x \in \mathbb{R}^n : x \geq 0, Ax \geq b, \langle c, x\rangle \leq \alpha\}$ is empty. Using Theorem A.1, its "dual set" $E_2(\alpha)$ is non empty. But this implies the non emptiness of $E_2(\alpha+\varepsilon)$ as well, for $\varepsilon > 0$ small enough, hence again by Theorem A.1 the emptiness of $E_1(\alpha+\varepsilon)$, contradicting the definition of α.
Let then x^* be a solution of $(\mathcal{P}_1)$. Thus $\{Ax \geq b, x \geq 0\}$ implies $\langle c, x\rangle \geq \langle c, x^*\rangle$. Using Corollary A.3 one obtains the existence of $(u, v) \in \mathbb{R}^m \times \mathbb{R}^n$ with:

$$(u, v) \geq 0, \quad uA + v = c, \quad \langle u, b\rangle \geq \langle c, x^*\rangle$$

hence u is an optimal solution of $\mathcal{P}_2$ and both programs have the same value. ∎

We are now in position to prove the celebrated Minmax Theorem obtained by Von Neumann (1928). $\Delta(I)$ denotes the simplex on I.

Theorem A.5
Let A be a $I \times J$ real matrix.
There exists a triple $(x^*, y^*, v) \in \Delta(I) \times \Delta(J) \times \mathbb{R}$ such that:

$$x^* A y \geq v, \quad \forall y \in \Delta(J)$$

$$x A y^* \leq v, \quad \forall x \in \Delta(I).$$

Proof
Adding a constant to all entries one can and will assume $a_{ij} > 0$ for all i, j. Denote by $\mathbf{1}_I$ the vector of ones in $\mathbb{R}^I$ and similarly for $\mathbf{1}_J$ in $\mathbb{R}^J$. Consider the following linear programs:

$$(\mathcal{Q}_1) \quad \begin{array}{l} \min \ \langle \mathbf{1}_I, x\rangle \\ xA \geq \mathbf{1}_J \\ x \geq 0 \end{array} \qquad\qquad (\mathcal{Q}_2) \quad \begin{array}{l} \max \ \langle y, \mathbf{1}_J\rangle \\ Ay \leq \mathbf{1}_I \\ y \geq 0 \end{array}$$

Note that they are both feasible (for $(\mathcal{Q}_1)$, A is strictly positive; for $(\mathcal{Q}_2)$ take $y = 0$) hence by Theorem A.4 they have the same value $w > 0$, and optimal solutions, say $\overline{x}$ and $\overline{y}$. It is clear that by renormalizing ($x^* = \overline{x}/w, y^* = \overline{y}/w$) one obtains solutions of the original problem with value $v = 1/w$. ∎

Von Neumann theorem says that any matrix game A has a value $v = \mathtt{val}A$ in mixed strategies :

$$v = \max_{x \in \Delta(I)} \min_{y \in \Delta(J)} xAy = \min_{y \in \Delta(J)} \max_{x \in \Delta(I)} xAy$$

and the players have optimal strategies.

Exercise 1 (Loomis, 1946)
Let A and B be two $I \times J$ matrices, with $B \gg 0$. There exists $v \in \mathbb{R}$, $x \in \Delta(I)$, $y \in \Delta(J)$ such that:

$$xA \geq v\, xB, \quad Ay \leq v\, By.$$

Hint
First the set of λ and x satisfying:

$$xA \geq \lambda\, xB$$

is nonempty and the upper bound λ_0 of such λ's is reached for some x^0. Similarly for the system

$$Ay \leq \mu\, By$$

with lower bound μ_0 and then necessarily $\lambda_0 \leq \mu_0$.
The proof of equality is by induction on $\#I + \#J$.
If both x^0 and y^0 are equalizing then $\lambda_0 = \mu_0$. So assume $J' \neq J$ with:

$$x^0 A_j > \lambda_0\, x^0 B_j, \quad \forall j \in J \setminus J'$$

$$x^0 A_j = \lambda_0\, x^0 B_j, \quad \forall j \in J'.$$

By induction let v' be associated to the $I \times J'$ sub-matrices of A and B. Obviously $v' \geq \mu_0$ and $v' \geq \lambda_0$. If $v' > \lambda_0$, choose x' optimal for the subsystem so that:

$$x' A_j \geq v'\, x' B_j > \lambda_0\, x' B_j, \quad \forall j \in J'.$$

For $\alpha > 0$ small enough one has:

$$[(1-\alpha)x^0 + \alpha x']A \gg \lambda_0[(1-\alpha)x^0 + \alpha x']B$$

contradicting the definition of λ_0. Hence $\lambda_0 = v' \geq \mu_0$ and the equality.

If $B_{ij} = 1$ for all i, j, one obtains the minmax theorem.
If $\#I = \#J$ and A is the identity, one obtains $v > 0$; then x and y both $\gg 0$ are eigenvectors of B and of its transpose (Perron Frobenius).

Exercice 2 (Ville, 1938)
Deduce from Theorem A.5 the following minmax theorem on the square.
Let $X = Y = [0, 1]$ and f be continuous on $X \times Y$. The game on $\Delta(X) \times \Delta(Y)$

has a value.
Take successive discrete dyadic approximations of the game.
Define the matrix game $A^n_{ij} = f(\frac{i}{2^n}, \frac{j}{2^n})$, $i, j = 0, ..., 2^n$, with value v_n and optimal strategy σ_n for Player 1, considered as a probability measure on $[0, 1]$. Prove that Player 1 can obtain $\limsup_{n\to\infty} v_n$ in the game f and that any accumulation point of the sequence $\{\sigma_n\}$ is an optimal strategy.

A.3 Sion's theorem and applications

The proof of Sion's Theorem will use the following result known as the Intersection Lemma (Berge, 1965):

Lemma A. 6
Let $X_i, i = 1, \ldots, n$, be a finite family of compact convex subsets of a locally convex Hausdorff topological vector space.
Assume that $X = \bigcup_{i=1}^n X_i$ is convex and that for any i, $\bigcap_{j,j\neq i} X_j \neq \emptyset$.
Then $\bigcap_i X_i \neq \emptyset$.

Proof
The proof goes by induction on the number of sets and clearly holds for $n = 2$ (See also the next part of the proof). Assuming the result false for n, the two convex and compact subsets $\bigcap_{i=1}^{n-1} X_i$ and X_n can be strongly separated (Hahn Banach) by a closed hyperplane H.
Introduce $X'_i = H \cap X_i$ and $X' = H \cap X$. These sets are compact convex, $X'_n = \emptyset$ and $X' = \bigcup_{i=1}^{n-1} X'_i$. Moreover, for any $i = 1, \ldots, n-1$, $\bigcap_{j\neq i,n} X'_j \neq \emptyset$. In fact the convex set $\bigcap_{j\neq i,n} X_j$ has by hypothesis a non empty intersection with both $\bigcap_{j\neq n} X_j$ and X_n that are separated by H. Hence its intersection with H is non empty.
The induction hypothesis now implies $\bigcap_{j\neq n} X'_j = H \cap \bigcap_{j\neq n} X_j \neq \emptyset$ a contradiction. ∎

The next theorem is due to Sion (1958).

Theorem A. 7
Let X and Y be convex subsets of topological vector spaces, one being compact. Assume that, for any real α and any couple (x_0, y_0) in $X \times Y$ the sets $\{x \in X; g(x, y_0) \geq \alpha\}$ and $\{y \in Y; g(x_0, y) \leq \alpha\}$ are closed and convex.
Then the game has a value (and there exists an optimal strategy on the compact side).

Proof
Assume, by contradiction, the existence of a real α with $\underline{v} < \alpha < \overline{v}$ and suppose, for instance, X compact. The family of sets $A_y = \{x \in X; g(x, y) < \alpha\}$

forms an open covering of X, hence there exists a finite set $Y_J = \{y_1, ..., y_n\}$ such that the family A_{y_j} covers X. Replacing Y by $Co(Y_J)$ preserves the conditions of the Theorem and the contradiction still holds. One can now use the compactness of Y and similarly introduce the open sets $B_x = \{y \in Co(Y_J);$ $g(x,y) > \alpha\}$ so that X reduces also to the convex hull of a finite set X_I of points. One can moreover assume that these sets are minimal for the property:

$$\forall x \in Co(X_I), \exists y_j \in Y_J, \quad g(x, y_j) < \alpha$$

$$\forall y \in Co(Y_J), \exists x_i \in X_I, \quad g(x_i, y) > \alpha$$

For each x_i, introduce $Y_i = \{y \in Co(Y_J); g(x_i, y) \leq \alpha\}$. These sets are compact and convex and have an empty intersection. Moreover by the minimality requirement $\bigcap_{i \neq i'} Y_i$ is nonempty for each i'. Hence by Lemma A.6 the union of the Y_i is not convex, in particular does not cover $Co(Y_J)$. Thus there exists $y^* \in Co(Y_J)$ with $g(x_i, y^*) > \alpha$ for all x_i, hence $g(., y^*) > \alpha$ on $Co(X_I)$ by quasiconcavity. Similarly there exists $x^* \in Co(X_I)$ with $g(x^*, .) < \alpha$ on $Co(Y_J)$, hence the contradiction by evaluating $g(x^*, y^*)$. ∎

Remarks

Sion's Theorem relies on properties involving the variables separately: $g(., y)$ is quasiconcave and upper semi continuous (u.s.c.), $g(x, .)$ quasiconvex and lower semi continuous (l.s.c.).

Only a finite dimensional version of the intersection lemma is used, hence the result uses only a finite dimensional version of the separation theorem.

Adding a geometrical hypothesis allows to weaken the topological assumptions as follows:

Proposition A.8

Let X be a compact convex subset of a topological vector space and Y be a convex set. Assume that for any real α and any y_0 in Y, the set $\{x \in X; g(x, y_0) \geq \alpha\}$ is closed and convex. Moreover for any $x_0 \in X$, $g(x_0, .)$ is convex on Y. Then the game on $X \times Y$ has a value.

Proof

As in the proof of Theorem A.7 one reduces the problem to the case where Y is the convex hull of a finite family $Y_J = \{y_j\}$ and the open sets A_{y_j} cover X. If y_j^n is a sequence of interior points of $Co(Y_J)$ converging to y_j, one has by convexity $\limsup_{n \to \infty} g(x, y_j^n) \leq g(x, y_j)$ for any $x \in X$. Thus the family of open sets $A_{y_j^n}$ also covers X and one can replace Y by the convex hull of finitely many interior points. On this set, all convex functions $g(x_0, .)$ are continuous and Theorem A.7 applies. ∎

A.4 Convexity and separation

This section covers results dealing with mixed strategies with finite support. They rely directly on a finite dimensional separation theorem.

Theorem A.9
Let X be a measurable space, Σ be a convex set of probability measures on X and J be finite. Let g be measurable and bounded on $X \times J$. Then the mixed game $(g; \Sigma, \Delta(J))$ has a value.

Proof
Let $g(\sigma, j) = \int_X g(x,j)\sigma(dx)$, $\underline{v} = \sup_\sigma \inf_j g(\sigma, j)$ and $D = \{a \in I\!R^J;\ \exists \sigma \in \Sigma,\ g(\sigma, j) = a_j, \forall j \in J\}$. Note that D is convex and has, for any $\varepsilon > 0$, an empty intersection with the convex set $D' = \{a \in I\!R^J; a_j \geq \underline{v} + \varepsilon,\ \forall j \in J\}$. Thus by the separation theorem, there exists $b \in I\!R^J$, $b \neq 0$ such that:

$$\langle a, b\rangle \leq \langle a', b\rangle, \quad \forall a \in D, \forall a' \in D'.$$

D' is upper comprehensive hence $b > 0$, so that by normalization there exists $y \in \Delta(J)$ with:

$$g(\sigma, y) = \sum\nolimits_j \int_X g(x,j)\sigma(dx) y_j \leq \underline{v} + \varepsilon, \quad \forall \sigma \in \Sigma.$$

Letting $\varepsilon \to 0$, one obtains by compactness a mixed strategy y with $g(\sigma, y) \leq \underline{v}, \forall \sigma \in \Sigma$, hence $\inf_y \sup_\sigma (g(\sigma, y) = \underline{v} = \overline{v}$ (and Player 2 has optimal strategy).

∎

Remark
Theorem A.9 follows also from Proposition A.8.

Adding topological hypotheses on X and g, one can relax the finiteness hypothesis on J.

Proposition A.10
Let X be compact, Y any set and g u.s.c. in X. The game on $\Delta(X) \times \Delta_f(Y)$ has a value and Player 1 has an optimal strategy.

Proof
For any finite subset $Y' \subset Y$ the set

$$C_n(Y') = \{\sigma \in \Delta(X); g(\sigma, y) \geq \inf_{\tau \in \Delta_f(Y)} \sup_{\sigma \in \Delta(X)} g(\sigma, \tau) - 1/n, \forall y \in Y'\}$$

is non empty by the previous Theorem A.9. The same property holds for any finite intersection of such sets. Since $g(., y)$ is u.s.c. on X for any y in Y, $g(., \tau)$

is u.s.c. on $\Delta(X)$ for τ in $\Delta_f(Y)$. The compactness of X implies that $\Delta(X)$ is compact hence the $C_n(Y')$ are also compact. So that their intersection for all n and all Y' is nonempty and any σ in it will satisfy:

$$g(\sigma, y) \geq \inf_{\tau \in \Delta_f(Y)} \sup_{\sigma \in \Delta(X)} g(\sigma, \tau), \qquad \forall y \in Y.$$

■

Remark
Proposition A.10 follows also directly from Proposition A.8.

A.5 Mixed extension

The next result, which is the "classical" minmax theorem for mixed strategies relies on Fubini's Theorem.

Theorem A. 11
Let X and Y be compact, g be bounded from below or from above and Borel measurable on $X \times Y$. Assume moreover that for any $(x_0, y_0) \in X \times Y$, $g(x_0, .)$ is l.s.c. on Y and $g(., y_0)$ u.s.c. on X. Then:

$$\sup_{\sigma \in \Delta_f(X)} \inf_{y \in Y} \int_X g(x, y)\sigma(dx) = \inf_{\tau \in \Delta_f(Y)} \sup_{x \in X} \int_Y g(x, y)\tau(dy)$$

The game on $\Delta(X) \times \Delta(Y)$ has a value, each player has an optimal strategy and for any $\varepsilon > 0$, an ε-optimal strategy with finite support.

Proof
We apply Proposition A.10 to the games $(g; \Delta(X), \Delta_f(Y))$ and $(g; \Delta_f(X), \Delta(Y))$ yielding to two values v^+ and v^- with obviously $v^+ \geq v^-$.
Let σ (resp. τ) be optimal for Player 1 (resp. 2) in the first (resp. second) game. Thus $\int_X g(x, y)\sigma(dx) \geq v^+, \forall y \in Y$ and $\int_Y g(x, y)\tau(dy) \leq v^-, \forall x \in X$. Using Fubini's Theorem one obtains:

$$v^+ \leq \int\int_{X \times Y} g(x, y)\sigma(dx)\tau(dy) \leq v^-$$

hence the result. ■

A.6 Purification

Definition
A function g on $X \times Y$ is **concave-like convex-like** if for any α, $0 < \alpha < 1$, and any couple $(x_1, x_2) \in X^2$ (resp. $(y_1, y_2) \in Y^2$) there exits $x_0 \in X$

(resp. $y_0 \in Y$) with: $g(x_0,.) \geq \alpha g(x_1,.) + (1-\alpha)g(x_2,.)$ on Y (resp. $g(.,y_0) \leq \alpha g(.,y_1) + (1-\alpha)g(.,y_2)$ on X).

The next result gives condition for "purification" to hold: mixed strategies are dominated by pure ones.

Proposition A. 12
Let g be concave like on $X \times Y$.
For any $\sigma \in \Delta_f(X)$ there exists $x \in X$ such that $g(x,.) \geq g(\sigma,.)$ on Y.
Assume X compact and $g(.,y)$ u.s.c. on X for any y. The same property holds for $\sigma \in \Delta(X)$.

Proof
For $\sigma \in \Delta_f(X)$ the proof is by induction on the support.
In the compact case, let Y_0 be a finite subset of Y. For any $\varepsilon > 0$, the law of large numbers applied to the (vector-valued) random variable $\{g(\zeta, y)\}_{y \in Y_0}$ where ζ follows the distribution σ on X implies the existence of $\sigma^{\varepsilon}_{Y_0} \in \Delta_f(X)$ satisfying: $g(\sigma^{\varepsilon}_{Y_0}, y) \geq g(\sigma, y) - \varepsilon$, $\forall y \in Y_0$. As above, $\sigma^{\varepsilon}_{Y_0}$ is dominated by a point in X. Hence the set $X^{\varepsilon}_{Y_0} = \{x \in X; g(x,y) \geq g(\sigma,y) - \varepsilon,\ \forall y \in Y_0\}$ is non empty. Moreover it is compact and the same argument shows that any finite collection of such sets has a non empty intersection. Thus the intersection of all of them is non empty and $x \in \cap_{\varepsilon > 0, Y_0 \text{finite}} X^{\varepsilon}_{Y_0}$ satisfies the requirement.
∎

The next result is due to Fan (1953).

Proposition A. 13
Let X be compact, Y any set and $g(.,y)$ u.s.c. on X for all $y \in Y$. Assume g concave-like and convex-like, then the game has a value on $X \times Y$.

Proof
Apply Proposition A.10 to the game $(g; \Delta(X), \Delta_f(Y))$, then purify using Proposition A.12.
∎

Remark
Note that Proposition A. 13 implies Proposition A.10 as well.

A.7 The value operator and the derived game

Let $\mathcal{F}$ be a convex cône of real functions on $X \times Y$ such that for each $f \in \mathcal{F}$ the game $(f; X, Y)$ has a value $\mathtt{val}_{X \times Y}(f) = \mathtt{val}(f)$.
Note that the operator $\mathtt{val}$:
is monotonic: $f \leq g$ implies $\mathtt{val}(f) \leq \mathtt{val}(g)$ and

reduces the constants: $\forall t \geq 0$, $\mathtt{val}(f+t) \leq \mathtt{val}(f) + t$. (In fact $\mathtt{val}(f+t) = \mathtt{val}(f) + t$ for all real t).
Hence we obtain:

Proposition A.14
The operator val *satisfies:*

$$|\mathtt{val}(f) - \mathtt{val}(g)| \leq \|f - g\|$$

with $\|f\| = \sup_{X\times Y} |f(x,y)|$.

Proof
Since $f \leq g + \|f-g\|$ the result follows from the two properties of the operator val. ∎

We consider now perturbations of a game and the corresponding directional derivatives of the value. This approach and the basic result in the finite case are due to Mills (1956).

Proposition A. 15
Let X and Y be compact sets, f and g real functions on $X \times Y$. Assume that for any $\alpha \geq 0$, the functions g and $f + \alpha g$ are u.s.c. in x and l.s.c. in y and that the game $(f + \alpha g; X, Y)$ has a value, $\mathtt{val}_{X\times Y}(f + \alpha g)$. *Then*

$$\mathtt{val}_{X(f)\times Y(f)}(g) = \lim_{\alpha \to 0+} \frac{\mathtt{val}_{X\times Y}(f + \alpha g) - \mathtt{val}_{X\times Y}(f)}{\alpha}$$

where $X(f)$ and $Y(f)$ are the optimal strategy sets in the game f.

Proof
The hypotheses imply that the sets of optimal strategies in the game $f + \alpha g$, $\alpha \geq 0$, $X(f + \alpha g)$ and $Y(f + \alpha g)$ are non empty. Let x_α in $X(f + \alpha g)$ and y in $Y(f)$. Then

$$\alpha g(x_\alpha, y) = (f + \alpha g)(x_\alpha, y) - f(x_\alpha, y) \geq \mathtt{val}_{X\times Y}(f + \alpha g) - \mathtt{val}_{X\times Y}(f).$$

So that:

$$\inf_{Y(f)} g(x_\alpha, y) \geq \frac{\mathtt{val}_{X\times Y}(f + \alpha g) - \mathtt{val}_{X\times Y}(f)}{\alpha}$$

and

$$\limsup_{\alpha \to 0+} \inf_{Y(f)} g(x_\alpha, y) \geq \limsup_{\alpha \to 0+} \frac{\mathtt{val}_{X\times Y}(f + \alpha g) - \mathtt{val}_{X\times Y}(f)}{\alpha}.$$

Let x be an accumulation point of a family $\{x_\alpha\}$, as α goes to 0 along a sequence realizing the lim sup. Since g is u.s.c. in x:

$$\inf_{Y(f)} g(x,y) \geq \limsup_{\alpha \to 0+} \inf_{Y(f)} g(x_\alpha, y) \ .$$

Note that x is in $X(f)$ (X is compact and $f + \alpha g$ is u.s.c. in x), hence:

$$\sup_{X(f)} \inf_{Y(f)} g(x,y) \geq \limsup_{\alpha \to 0+} \frac{\mathtt{val}_{X\times Y}(f+\alpha g) - \mathtt{val}_{X\times Y}(f)}{\alpha}$$

and the result follows from the dual inequality. ∎

The game on $X(f)\times Y(f)$ with payoff g is called the **derived game** of f in the direction g.

A.8 Fenchel's duality

We first introduce some notations.
co C (resp. $\overline{co}\ C$) denotes the convex (rep. closed convex) hull of a set C: the smallest convex (resp. closed convex) containing C. Let f be a function defined on $X = \mathbb{R}^n$ with values in $\mathbb{R}\cup\{+\infty\}$. Its **epigraph** is $\mathtt{epi}(f) = \{(x,t)\in X\times\mathbb{R}; f(x) \leq t\}$. The projection of $\mathtt{epi}(f)$ on X is its **domain** $\mathtt{Dom}(f) = \{x\in X : f(x) < +\infty\}$. f will always be assumed **proper** : $\mathtt{Dom}(f)\neq\emptyset$. f is l.s.c. iff its $\mathtt{epi}(f)$ is closed.
The **Fenchel conjugate** of f is defined on $\mathbb{R}^n$ by:

$$f^*(p) = \sup_{x\in\mathbb{R}^n} \{\langle x,p\rangle - f(x)\}$$

Note that f^* is convex and l.s.c.. The celebrated Fenchel's theorem is:

Theorem A. 16
Assume f convex and l.s.c.. Then:

$$f = f^{**}$$

Proof
By definition:

$$f(x) \geq \langle x,p\rangle - f^*(p)$$

so that

$$f \geq f^{**}.$$

Assume now: $a < f(x) < +\infty$. The separation theorem applied to the compact convex set (x,a) and the closed convex set $\mathtt{epi}(f)$ gives the existence of $\ell\in\mathbb{R}^n$ and $u\in\mathbb{R}$ with:

$$\langle \ell,x\rangle + ua < t < \langle \ell,y\rangle + ub, \quad \forall (y,b)\in\mathtt{epi}(f).$$

Taking $y = x$ above implies $u > 0$, hence one can choose $u = 1$ and $t - \langle \ell, . \rangle$ is an affine minorant of f. In particular $f^*(-\ell) < -t < -\langle \ell, x\rangle - a$. Finally one obtains $f^{**}(x) \geq \langle -\ell, x\rangle - f^*(-\ell) > a$.
Hence, $f(x) < +\infty$ implies $f(x) = f^{**}(x)$.
For the case $a < f(x) = +\infty$, we proceed as before. If $u \neq 0$ the same proof applies. Otherwise one has $u = 0$, hence:

$$\langle \ell, x\rangle < t < \langle \ell, y\rangle, \quad \forall y \in \mathtt{Dom}(f).$$

Since f is proper, by the previous analysis there exists an affine minorant to f, say $\langle \alpha, .\rangle + \beta$. Hence for $\theta > 0$ we obtain a new affine minorant by:

$$\langle \alpha, y\rangle + \beta + \theta(t - \langle \ell, y\rangle) \leq f(y)$$

while for θ large enough its value at x exceeds a. So that $f^{**} > a$ again.

∎

Remark
The previous proof shows that f^{**} is the largest convex, l.s.c. function below f: $\mathtt{epi}(f^{**}) = \overline{co}\ \mathtt{epi}(f)$.

Definition
The **subdifferential** of f at x is defined by:

$$\partial f(x) = \{p \in \mathbb{R}^n : f(y) \geq f(x) + \langle y - x, p\rangle, \forall y \in \mathbb{R}^n\}$$

Proposition A.17

$$p \in \partial f(x) \iff f(x) + f^*(p) = \langle x, p\rangle.$$

Proof
By definition, the next inequality holds:

$$f(y) + f^*(q) \geq \langle y, q\rangle \qquad \forall y, q \in \mathbb{R}^n \times \mathbb{R}^n.$$

Now one has:

$$\begin{aligned} p \in \partial f(x) &\iff f(y) \geq f(x) + \langle y - x, p\rangle, \quad \forall y \in \mathbb{R}^n \\ &\iff \langle x, p\rangle - f(x) \geq \langle y, p\rangle - f(y), \quad \forall y \in \mathbb{R}^n \\ &\iff \langle x, p\rangle - f(x) \geq f^*(p) \end{aligned}$$

which gives the required equality. ∎

In all following results, we deal with a real function f defined on a compact convex subset $K \subset X$, extended by $+\infty$ outside K.
Let $\mathtt{Vex}f$ be the largest function, convex and below f on X. Then one has: $\mathtt{Vex}f(x) = \inf\{t; (x,t) \in co\ \mathtt{epi}(f)\}$. Two basic properties are the following.

Proposition A.18
Assume f l.s.c. and bounded on K. Then $\mathtt{epiVex}\ f = \overline{co}\ \mathtt{epi} f = co\ \mathtt{epi} f$ *and* $\mathtt{Vex} f$ *is l.s.c..* $\mathtt{Vex}$ *coincides on K with the largest function convex and below f on K.*
Assume f convex and K be a polytope. Then f is u.s.c..

The next condition does not assume f convex.

Proposition A.19
Let f be l.s.c. and bounded on K. Assume $\partial f^*(p) = \{x\}$, then:

$$f(x) + f^*(p) = \langle x, p\rangle$$

Proof
The previous Proposition applied to f^* proves that x is the only solution of:

$$f^{**}(x) + f^*(p) = \langle x, p\rangle$$

In particular x is an extreme point of f^{**} in the sense that a decomposition $x = \sum \lambda_i x_i$ with I finite, $\lambda \in \Delta(I)$ and $f^{**}(x) = \sum_i \lambda_i f^{**}(x_i)$ would imply $x = x_i, \forall i \in I$.
The conditions on f then gives $f^{**} = \mathtt{Vex} f$ and x is an extreme point of $co\ \mathtt{epi} f$ hence $\mathtt{Vex} f(x) = f(x)$ ■

In our framework of incomplete information games the initial function u will be continuous on a simplex $\Delta(K)$. One extend it by $-\infty$ elsewhere. Consider the two operators Λ_s and Λ_i:

$$\Lambda_s(f)(x) = \sup_{y \in \mathbb{R}^n} \{f(y) - \langle x, y\rangle\}$$

and

$$\Lambda_i(f)(y) = \inf_{x \in \mathbb{R}^n} \{f(x) + \langle x, y\rangle\}.$$

It then follows from Fenchel's theorem and Proposition A.18 that:

$$\Lambda_i \circ \Lambda_s(u) = \mathtt{Cav}\ (u)$$

where $\mathtt{Cav}\ u$ is the smallest function concave and greater than g on $\Delta(K)$ (and $-\infty$ elsewhere).
Note that, by Proposition A. 18, $\mathtt{Cav}\ u$ is continuous on $\Delta(K)$.

A.9 Comments

Most of the material of this Appendix is taken from MSZ, Chapter I.
Proposition A. 18 follows Rockafellar (1970), Theorem 10.2 p.84 and Corollary 17.2.1 p.158.

B: Approachability theory

All the results in this section, except when indicated, are due to Blackwell (1956). Some of the proofs are taken from MSZ (Chapter II, Section 4).

The framework is as follows:
A is a $I\times J$ matrix with coefficients in $\mathbb{R}^K$. At each stage n, Player 1 (resp. Player 2) chooses a move, i_n in I (resp. j_n in J). The corresponding vector payoff, $g_n = A_{i_n j_n}$ is then announced. Denote by h_n the sequence of payoffs before stage n (this is, at least, the information available to both players at stage n) and let $\overline{g}_n = \frac{1}{n}\sum_{m=1}^{n} g_m$ be the average payoff up to stage n.
Let also $\|A\| = \max_{i\in I, j\in J, k\in K} |A^k_{ij}|$

Definitions
A set C in $\mathbb{R}^K$ is **approachable** by Player 1 if for any $\varepsilon > 0$ there exists a strategy σ and N such that, for any strategy τ of Player 2 and any $n \geq N$:

$$E_{\sigma,\tau}(d_n) \leq \varepsilon$$

where d_n is the euclidean distance $d(\overline{g}_n, C)$.
A set C in $\mathbb{R}^K$ is **excludable** by Player 1 if for some $\delta > 0$, the set $C^c_\delta = \{z; d(z, C) \geq \delta\}$ is approachable by him.
A dual definition holds for Player 2.
Notice that if Player 1 can approach C, Player 2 cannot exclude C, he could however approach C^c. The analog of the minmax theorem would be: if Player 1 cannot approach C then Player 2 can exclude it. The next analysis will prove this to be true within the class of convex set and a counter example will be given for general set C.
From the definitions it is enough to consider closed sets C and even their intersection with the closed ball of radius $\|A\|$.
Given s in $S = \Delta(I)$, define $sA = co\,\{\sum_i s_i A_{ij}; j\in J\}$, and similarly At, for t in $T = \Delta(J)$. If Player 1 uses s his expected payoff will be in sA.
The first result is a sufficient condition for approachability based on the following notion:

Definition
A closed set C in $\mathbb{R}^K$ is a **B**-set for Player 1 if: for any $z\notin C$, there exists a

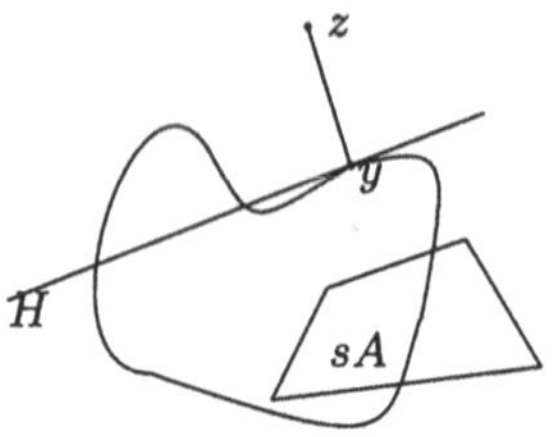

Fig. B.1.

closest point $y = y(z)$ in C to z and a mixed move $s = \mathbf{s}(z)$ in S, such that the hyperplane trough y orthogonal to the segment $[yz]$ separates z from sA. Note that for any $x \in sA$, any point on the line $[xz]$ closed to z will be closer to y than z itself. This geometric consideration is the basis of the next result.

Theorem B1

Let C be a **B***-set for Player 1. Then C is approachable by that player. More precisely with a strategy satisfying $\sigma(h_{n+1}) = \mathbf{s}(\overline{g}_n)$, whenever $\overline{g}_n \notin C$, one has:*

$$E_{\sigma\tau}(d_n) \;\leq\; \frac{2\|A\|}{\sqrt{n}} \qquad \forall \tau$$

and d_n converges $P_{\sigma\tau}$ a.s. to 0.

Proof

Let Player 1 use a strategy σ as above. Denote $y_n = y(\overline{g}_n)$. Then one has:

$$\begin{aligned} d_{n+1}^2 &\leq \|\overline{g}_{n+1} - y_n\|^2 \\ &= \|\frac{1}{n+1}(g_{n+1} - y_n) + \frac{n}{n+1}(\overline{g}_n - y_n)\|^2 \\ &= (\frac{1}{n+1})^2\|(g_{n+1} - y_n)\|^2 + (\frac{n}{n+1})^2 d_n^2 + \frac{2n}{(n+1)^2}\langle g_{n+1} - y_n, \overline{g}_n - y_n\rangle. \end{aligned}$$

The property of $\mathbf{s}(\overline{g}_n)$ implies that:

$$E(\langle g_{n+1} - y_n, \overline{g}_n - y_n\rangle | h_{n+1}) \leq 0$$

since $E(g_{n+1}|h_{n+1})$ belongs to $\mathbf{s}(\overline{g}_n)A$.

We can assume C included in the ball of radius $\|A\|$, hence $E(\|(g_{n+1} - y_n)\|^2 | h_{n+1}) \leq 4\|A\|^2$ and

$$E(d_{n+1}^2 | h_{n+1}) \leq \frac{4\|A\|^2}{(n+1)^2} + (\frac{n}{n+1})^2 d_n^2 \tag{B.1}$$

which implies by induction:

$$E(d_n^2) \le \frac{4\|A\|^2}{n}.$$

To get almost sure convergence, let $e_n = d_n^2 + \sum_{m>n} \frac{4\|A\|^2}{m^2}$ so that, by (B.1): $E(e_{n+1}|h_{n+1}) \le e_n$. Thus e_n is a positive supermartingale that majorizes d_n^2 and satisfies $E(e_n) \le 4\|A\|^2(\frac{1}{n} + \sum_{m>n} \frac{1}{m^2})$. Hence it converges to 0 a.s. and d_n also. ∎

Corollary B2

For any s in S, sA is approachable by Player 1 with the constant strategy s.

It follows that a necessary condition for a set C to be approachable by Player 1 is that for any t in T, $At \cap C \neq \emptyset$, otherwise C would be excludable by Player 2. In fact this condition is also sufficient for convex sets.

Theorem B3

Assume C closed and convex in $\mathbb{R}^K$. C is a **B**-*set for Player 1 iff $At \cap C \neq \emptyset$, for all t in T.*
In particular a set is approachable iff it is a **B**-*set.*

Proof

By the previous Corollary B2, it is enough to show that if $At \cap C \neq \emptyset$ for all t, C is a **B**-set.

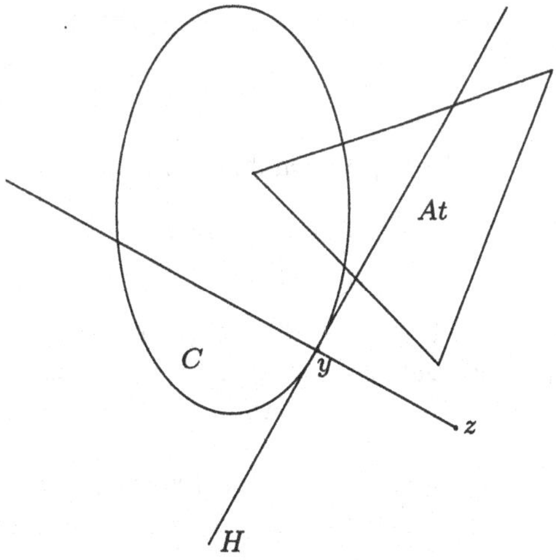

Fig. B.2.

The idea is to reduce the problem to the one dimensional case and to use the minmax theorem.

In fact, let $z \notin C$, y be its projection on C and consider the game with real payoff $\langle y-z, A\rangle$. Since $At \cap C \neq \emptyset$ for all t, it implies that its value is at least $\langle y-z, y\rangle = \min_{c \in C} \langle y-z, c\rangle$. Hence there exists an optimal strategy s for Player 1 such that $\langle y-z, \sum_i s_i A_{ij}\rangle \geq \langle y-z, y\rangle$ for any j in J, which shows that sA is on the opposite side of the hyperplane and the result follows. ∎

The previous proof gives the following practical criteria:

Corollary B4
A closed convex set C is a **B***-set for Player 1 iff, for any* α *in* $\mathbb{R}^K$*:*

$$\mathtt{val}\langle \alpha, A\rangle \geq \inf_{c \in C} \langle \alpha, c\rangle.$$

In dimension 1, any set is either approachable or excludable. In fact, define v (resp. v') as the value of the zero-sum game where Player 1 maximizes (resp. minimizes): $v = \max_s \min_t sAt$ and $v' = \min_s \max_t sAt$, then one has:

Proposition B5
Case 1: $v' \leq v$. *C is a* **B***-set for Player 1 iff* $C \cap [v', v] \neq \emptyset$ *and excludable by Player 2 otherwise.*
Case 2: $v \leq v'$. *C is a* **B***-set for Player 1 iff C contains* $[v, v']$ *and excludable by Player 2 otherwise.*

Proof
For the first case, if $C \cap [v', v] \neq \emptyset$, C is clearly a **B**-set for Player 1. On the other hand $[v', v]$ is a **B**-set for Player 2 hence approachable by him.
The other case is similar.

∎

An example of a set neither approachable nor excludable is given by the following game:

$$A = \begin{pmatrix} (0,0) & (0,0) \\ (1,0) & (1,1) \end{pmatrix}$$

and the set $C = \{(1/2, y);\ 0 \leq y \leq 1/4\} \cup \{(1, y);\ 1/4 \leq y \leq 1\}$.
If Player 1 plays n times *Bottom* and then *Top* or *Bottom* for the next n stages depending whether $\overline{g}_n^2$ is greater than 1/2 or not, the average $\overline{g}_{2n}$ will be in C.
However Player 2 can prevent the average payoff of remaining near C by either playing Left (and reaching the horizontal axis) or Right (and reaching the diagonal).
This leads to the following notion of weak approachability.

Definitions
A set C in $\mathbb{R}^K$ is **weakly approachable** by Player 1 if for any $\varepsilon > 0$ there exists N such that for any $n \geq N$ there is a strategy $\sigma = \sigma(n)$ satisfying:

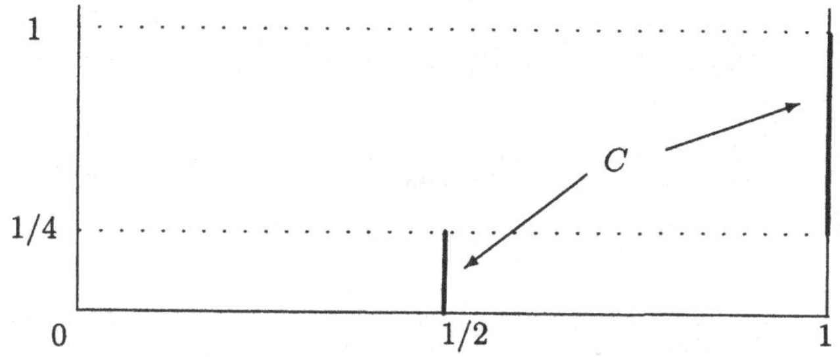

Fig. B.3.

for any strategy τ of Player 2, $E_{\sigma,\tau}(d_n) \leq \varepsilon$.
C is **weakly excludable** by Player 1 if for some $\delta > 0$, the set $C^c_\delta = \{z;\ d(z, C) \geq \delta\}$ is weakly approachable by him.

In the example above the described strategy of player 1 shows that C is weakly approachable.
In fact, the sequence of un-normalized cumulative payoffs, $(\frac{1}{n}\sum_{\ell \leq m} g_\ell)$, $m = 1, \ldots, n$ defines a piecewise linear curve and Player 1's objective is to reach C at stage n. It is thus natural to consider the game played in continuous time between 0 and 1 with position $\int_0^t g_u du$ at time t.
Again in the previous example, Player 1 can generate a curve with slope between 0 and 1 (controlled by Player 2) and horizontal speed 1 but he can also stop the game. Clearly he is thus able to reach C before time 1.
This result is general since one has:

Theorem B6 (Vieille, 1992)
Any set is either weakly approachable or weakly excludable.

Sketch of the proof
The proof uses the theory of differential games of fixed duration.
The idea is to consider v_n as the value of the discretisation of a differential game Γ played between time 0 and 1 and with payoff $\mathbf{1}_C(\int_0^1 g_u du)$.
Formally the deterministic dynamic is given by:

$$\frac{dz(t)}{dt} = \alpha(t) A \beta(t)$$

where α and β are the controls of the players with $\alpha(t) \in \Delta(I)$ and $\beta(t) \in \Delta(J)$ for $t \in [0, 1]$.
Take a smooth payoff function $g(z) = 1 - d(z, C)$.
For any initial point $z \in \mathbb{R}^K$ and time $\xi \in [0, 1]$, one defines two discretisations Γ_n^- and Γ_n^+ where the controls are constant on $(\frac{m}{n}(1-\xi), \frac{m+1}{n}(1-\xi))$, the payoff is $g(z(1))$ with $z(\xi) = z$ and Player 1 (resp. Player 2) is playing first.
Their values $W_n^-(\xi, z)$ and $W_n^+(\xi, z)$ satisfy:

$$W_n^-(\xi, z) = \max_{\alpha_1\in\Delta(I)} \min_{\beta_1\in\Delta(J)} \dots \max_{\alpha_n\in\Delta(I)} \min_{\beta_n\in\Delta(J)} g_\delta(z+(1-\xi)\frac{1}{n}\sum_{m=1}^{n} \alpha_m A\beta_m)$$

$$W_n^+(\xi, z) = \min_{\beta_1\in\Delta(J)} \max_{\alpha_1\in\Delta(I)} \dots \min_{\beta_n\in\Delta(J)} \max_{\alpha_n\in\Delta(I)} g_\delta(z+(1-\xi)\frac{1}{n}\sum_{m=1}^{n} \alpha_m A\beta_m)$$

Moreover the following recursive equations hold:

$$W_n^-(\xi, z) = \max_{\alpha\in\Delta(I)} \min_{\beta\in\Delta(J)} W_{n-1}^-(\xi + \frac{1}{n}(1-\xi), z + \frac{1}{n}(1-\xi)\alpha A\beta)$$

$$W_n^+(\xi, z) = \min_{\beta\in\Delta(J)} \max_{\alpha\in\Delta(I)} W_{n-1}^+(\xi + \frac{1}{n}(1-\xi), z + \frac{1}{n}(1-\xi)\alpha A\beta)$$

The main results we use are, see e.g. Souganidis (1999):
1) W_n^- and W_n^+ converge to some functions W^- and W^+ as n goes to ∞.
2) W^- is a viscosity solution on $[0,1]$ of the equation:

$$\frac{\partial U}{\partial t} + \max_{\alpha\in\Delta(I)} \min_{\beta\in\Delta(J)} \langle \nabla U, \alpha A\beta\rangle = 0$$

$$U(1, z) = g(z)$$

3) this solution is unique.
A similar result for W^- and the property:

$$\max_{\alpha\in\Delta(I)} \min_{\beta\in\Delta(J)} \langle \nabla U, \alpha A\beta\rangle = \min_{\beta\in\Delta(J)} \max_{\alpha\in\Delta(I)} \langle \nabla U, \alpha A\beta\rangle$$

finally imply:

$$W^- = W^+$$

and we denote this value by W.
Hence if $W(0,1) = 1$, for any $\varepsilon > 0$ there exists N such that if $n \geq N$ Player 1 can force an outcome within ε of C in Γ_n^-.
It remains to show that he can generate almost the same trajectory in the initial discrete time repeated game. Consider first G_{rn}. Inductively Player 1 will play by blocks of length r: on the m-th block he computes the optimal control α_m of stage m in Γ_n^- (given a past generated in the differential game by β_ℓ, $\ell < m$, which is the average behavior of Player 2 on block ℓ) and play the mixed move α_m i.i.d. for r stages. Use then Theorem B1 to prove that if d_m denotes the distance between $\alpha_m A\beta_m$ and the average payoff on block m in G_{rn}, it satisfies $E(d_m) \leq \frac{2\|A\|}{\sqrt{r}}$. It follows that for r large enough the average payoff in G_{rn} is within ε of one trajectory compatible with an optimal strategy in Γ_n^- hence within 2ε of C.
Finally if $m = rn + q$ the error in approximating G_m by G_{rn} is at most of the order $1/r$.
If $W(0,1) < 1$, Player 2 can force the outcome to belong to the complement of a δ neighborhood of C for δ small enough and a similar construction shows that C is weakly excludable. ∎

Remarks

The results extend to a much more general model with random payoffs and measurable move spaces, see Blackwell (1956), MSZ II.4 and to infinite dimensional spaces Lehrer (1997).

Spinat (2000) proved that a set is approachable iff it contains a **B**-set.
In particular it implies that our "light" definition of approachability (convergence in expectation) is in fact equivalent to a "heavy" one (almost sure convergence). It also shows that there are always robust approachability strategies (for example: independent of ε and of the previous own moves).
Note however that the a.s. convergence does not extend to a stochastic framework.

C: Operators and repeated games

C.1 Non-expansive mappings

This section follows Kohlberg and Neyman (1981).
Let Ψ be a **non-expansive mapping** on a Banach space X, i.e. such that:

$$\|\Psi(x) - \Psi(y)\| \leq \|x - y\| \qquad \forall x, y \in X.$$

The mapping $x \mapsto \Psi[(1-\lambda)x]$ is contracting for any λ in $(0,1)$. Denote by x_λ the corresponding fixed point:

$$x_\lambda = \Psi[(1-\lambda)x_\lambda].$$

We are interested in the asymptotic behavior of $\frac{\Psi^n(0)}{n}$ as $n \to \infty$ and of λx_λ as λ goes to 0. These quantities correspond to the values of finite and discounted games, see Section 4. Define:

$$\rho = \inf_{x \in X} \|\Psi(x) - x\|.$$

Theorem C.1 (Kohlberg and Neyman, 1981)
a) For all x in X:

$$\lim_{n\to\infty} \|\frac{\Psi^n(x)}{n}\| = \lim_{\lambda \to 0} \|\lambda x_\lambda\| = \rho.$$

b) Assume $\rho > 0$. For any x in X here exists ℓ_x in the unit ball B' of the dual of X such that,

$$\ell_x(\Psi^n(x) - x) \geq n\rho, \quad \forall n \geq 1$$

and

$$\lim_{\lambda \to 0} \ell_0(\lambda x_\lambda) = \rho.$$

The proof is divided in several steps.
First the next properties follow directly from the non-expansive aspect.

Lemma C.2

$$\lim_{n\to\infty} \|\frac{\Psi^n(x)}{n} - \frac{\Psi^n(y)}{n}\| = 0$$

and

$$\limsup_{n\to\infty} \|\frac{\Psi^n(x)}{n}\| \le \|\Psi(y) - y\|, \quad \forall x, y \in X.$$

Proof

In fact:

$$\|\frac{\Psi^n(x)}{n} - \frac{\Psi^n(y)}{n}\| \;\le\; \frac{1}{n}\|x - y\|$$

and thus $\limsup_{n\to\infty} \|\frac{\Psi^n(x)}{n}\|$ is independent of x. Moreover:

$$\begin{aligned}\|\Psi^n(y) - y\| &= \|\Psi^n(y) - \Psi^{n-1}(y) + \Psi^{n-1}(y) + \ldots - y\| \\ &\le n\|\Psi(y) - y\|\end{aligned}$$

so that:

$$\limsup_{n\to\infty} \|\frac{\Psi^n(y)}{n}\| \;\le\; \|\Psi(y) - y\|$$

∎

Define $y_\lambda = (1-\lambda)\, x_\lambda$ so that $\Psi(y_\lambda) = \frac{y_\lambda}{1-\lambda}$.

Lemma C.3

$$\frac{\lambda}{1-\lambda}\|y_\lambda\| \;\le\; 2\|\frac{\lambda}{1-\lambda}y\| + \|\Psi(y) - y\|, \quad \forall y \in X.$$

Proof

$$\begin{aligned}\|y_\lambda - y\| \ge \|\Psi(y_\lambda) - \Psi(y)\| &= \|\frac{y_\lambda}{1-\lambda} - \Psi(y)\| \\ &\ge \frac{1}{1-\lambda}\|y_\lambda - y\| - \|\frac{\lambda}{1-\lambda}y\| - \|\Psi(y) - y\|\end{aligned}$$

Hence

$$\frac{\lambda}{1-\lambda}\|y_\lambda - y\| \;\le\; \|\frac{\lambda}{1-\lambda}y\| + \|\Psi(y) - y\|$$

∎

Corollary C.4

$$\limsup_{\lambda\to 0} \|\lambda x_\lambda\| \;\le\; \|\Psi(y) - y\|, \quad \forall y \in X.$$

Corollary C.5

$$\lim_{\lambda\to 0} \|\lambda x_\lambda\| = \rho.$$

Proof

By definition:

$$\rho \leq \|\Psi(y_\lambda) - y_\lambda\| = \|\Psi((1-\lambda)x_\lambda) - (1-\lambda)x_\lambda\| = \|\lambda x_\lambda\|$$

and the result follows from Corollary C.4. ∎

From Lemma C.2 and Corollary C.4 one already deduces that, for any ℓ in B':

$$\limsup_{n\to\infty} \ell\left(\frac{\Psi^n(x)}{n}\right) \leq \limsup_{n\to\infty} \|\frac{\Psi^n(x)}{n}\| \leq \rho$$
$$\limsup_{\lambda\to 0} \ell(\lambda x_\lambda) \leq \limsup_{\lambda\to 0} \|\lambda x_\lambda\| \leq \rho$$

and that the last inequality to prove for a) ($\liminf_{n\to\infty} \|\frac{\Psi^n(x)}{n}\| \geq \rho$) will follow from b).

The geometric argument behind b) is as follows. Assume $\rho > 0$ and remark that, by Corollary C.5, $\|x_\lambda\|$ and $\|y_\lambda\|$ go to ∞. Hence for a given y, $\Psi(y_\lambda) - \Psi(y)$ and $y_\lambda - \Psi(y)$ are nearly parallel. Hence $\|\Psi(y_\lambda) - \Psi(y)\|$ is near $\|y_\lambda - \Psi(y)\| + \frac{\lambda}{1-\lambda}\|y_\lambda\|$, but on the other hand cannot exceed $\|y_\lambda - y\|$.

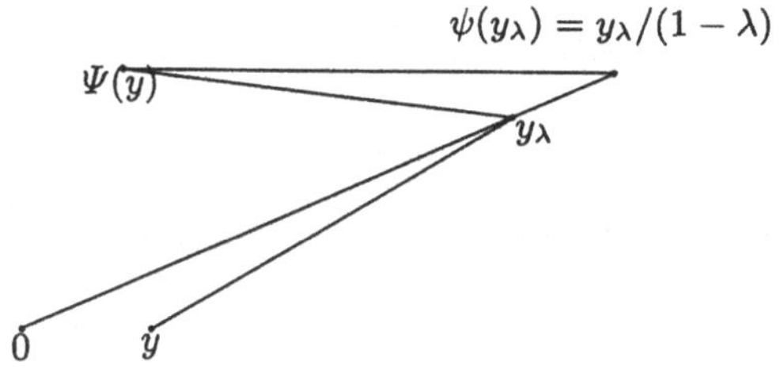

Fig. C.1.

Hence an application of Ψ pushes y towards y_λ by an amount of $\|\lambda x_\lambda\|$ which is at least ρ. Formally,

Lemma C.6

$$\|\Psi(y) - y_\lambda\| \leq \|y - y_\lambda\| - \rho + \frac{2\lambda}{1-\lambda}\|\Psi(y)\|.$$

Proof

$$\begin{aligned}\|\Psi(y) - y_\lambda\| &= \frac{1}{1-\lambda}\|\Psi(y) - y_\lambda\| - \frac{\lambda}{1-\lambda}\|\Psi(y) - y_\lambda\| \\ &\leq \|\Psi(y) - \frac{y_\lambda}{1-\lambda}\| + \frac{2\lambda}{1-\lambda}\|\Psi(y)\| - \frac{\lambda}{1-\lambda}\|y_\lambda\| \\ &\leq \|y - y_\lambda\| + \frac{2\lambda}{1-\lambda}\|\Psi(y)\| - \rho.\end{aligned}$$

∎

Lemma C.7
Given $y \neq 0$, let ℓ_y in the unit ball of X' with $\ell_y(y) = \|y\|$. Then: $\|y - x\| \le \|y\| - \beta$ implies $\ell_y(x) \ge \beta$.

Proof

$$\|y\| - \ell_y(x) = \ell_y(y - x) \le \|y - x\|.$$

∎

Applying Lemma C.6 n times starting from y gives:

$$\|\Psi^n(y) - y_\lambda\| \le \|y - y_\lambda\| - n\rho + \frac{2\lambda}{1-\lambda}\sum\nolimits_{m=1}^{n} \|\Psi^m(y)\|$$

Use then Lemma C.7 with $x = \Psi^n(y) - y$ and $\ell_y^\lambda = \ell_{y_\lambda - y}$ to deduce:

$$\ell_y^\lambda(\Psi^n(y) - y) \ge n\rho - \frac{2\lambda}{1-\lambda}\sum\nolimits_{m=1}^{n} \|\Psi^m(y)\|.$$

Let ℓ_y^* be an accumulation point in the weak* topology of the family ℓ_y^λ as λ goes to zero. Then

$$\ell_y^*(\Psi^n(y) - y) \ge n\rho, \quad \forall n.$$

Note that $0 \neq \|\ell_y^*\| \le 1$ and take $\ell_y = \ell_y^*/\|\ell_y^*\|$ if needed.

Finally from

$$\|\Psi(y_\lambda) - \Psi(y_\mu)\| \le \|y_\lambda - y_\mu\|$$

one obtains

$$\|y_\lambda - y_\mu\| \le \|y_\mu\| - \frac{\mu(1-\lambda)}{\lambda(1-\mu)}\|y_\mu\|$$

hence using Corollary C.5 and Lemma C.7:

$$\ell_0^\mu(y_\lambda) \ge \frac{1-\lambda}{\lambda}\rho$$

hence

$$\ell_0^\mu(\lambda x_\lambda) \ge \rho$$

so that as well:

$$\ell_0^*(\lambda x_\lambda) \ge \rho.$$

∎

This ends the proof of Theorem C.1.

C.2 Bounded variation

Given a non-expansive mapping Ψ on a Banach space, define v_λ and v_n by $v_0 = 0$ and:

$$v_\lambda = \lambda\Psi[\frac{(1-\lambda)}{\lambda}v_\lambda]$$

$$v_n = \frac{1}{n}\Psi[(n-1)v_{n-1}].$$

According to the previous notations one has $v_\lambda = \lambda x_\lambda$ and $v_n = \dfrac{\Psi^n(0)}{n}$.

Definition
v_λ is of **bounded variation** if for any decreasing sequence $\lambda_i \in (0,1)$:

$$\sum_i \|v_{\lambda_{i+1}} - v_{\lambda_i}\| < \infty.$$

In particular v_λ converges to some v.

Theorem C.8 (Neyman, 1998)
Assume v_λ of bounded variation. Then v_n converges to v.

Proof
Let us write w_n for v_λ when $\lambda = 1/n$. In particular w_n converges to v.

$$\begin{aligned}\|v_{n+1} - w_{n+1}\| &= \|\frac{1}{n+1}\Psi[nv_n] - \frac{1}{n+1}\Psi[nw_{n+1}]\| \\ &< (\frac{n}{n+1})\|v_n - w_{n+1}\| \\ &\leq (\frac{n}{n+1})(\|v_n - w_n\| + \|w_n - w_{n+1}\|).\end{aligned}$$

Hence by summation:

$$(n+1)\|v_{n+1} - w_{n+1}\| \leq \|v_1 - w_1\| + \sum_{m=1}^{n} m\ \|w_{m+1} - w_m\|.$$

Since $\sum_{m=1}^{\infty}\|w_{m+1} - w_m\|$ is bounded,

$$\lim_{n\to\infty}\frac{1}{n+1}\sum_{m=1}^{n} m\|w_{m+1} - w_m\| = 0$$

hence the result.

∎

Remarks
The same computation as above gives:

$$\|v_{n+1} - w_{n+1}\| \leq \frac{n}{n+1}\|v_n - w_{n+1}\| \leq \frac{n}{n+1}(\|v_n - v_{n+1}\| + \|v_{n+1} - w_{n+1}\|)$$

hence:

$$\|v_{n+1} - w_{n+1}\| \leq n\|v_{n+1} - v_n\|.$$

So that if $\|v_n - v\| = o(1/n)$, $\lim_{n\to\infty} v_{\lambda_n} = \lim_{n\to\infty} v_n$ on the subsequence $\lambda_n = 1/n$. But for $\lambda > \mu$:

$$\|x_\lambda - x_\mu\| \leq \frac{\lambda - \mu}{\lambda}\|x_\mu\|$$

hence $\lim_{\lambda\to 0} v_\lambda = v$.
Also, if v_n is of bounded variation, $\liminf_{n\to\infty} n\|v_{n+1} - v_n\|$ is 0, hence there exists a subsequence of v_λ converging to $\lim_{n\to\infty} v_n$ as $\lambda\to 0$.

C.3 Approximate fixed points

This section follows De Meyer and Rosenberg (1999).

C.3.1 Finite version

Consider a sequence of operators $\Psi_n, n \geq 1$, on a Banach space X and define:

$$v_1 = \Psi_1(0)$$

and

$$v_{n+1} = \Psi_{n+1}(v_n).$$

Assume that the family has the following contracting property:

$$\|\Psi_{n+1}(x) - \Psi_{n+1}(x')\| \leq (\frac{n}{n+1})^a \|x - x'\|$$

for some positive constant a, and n large enough.
For example, $\mathbf{\Psi}$ is non-expansive and $\Psi_{n+1}(x) = \frac{1}{n+1}\mathbf{\Psi}(nx)$.

Theorem C.9
If a sequence x_n satisfies, for some positive b, and n large enough:

$$\|\Psi_{n+1}(x_n) - x_{n+1}\| \leq \frac{1}{(n+1)^{1+b}}$$

and converges strongly to x, then v_n converges strongly to x also. Moreover, for any $0 < c < \min(a, b)$ there exists a constant A such that:

$$\|v_n - x_n\| \leq \frac{A}{n^c}\,.$$

Proof
Write, for $n \geq N$ large enough:

$$\|v_{n+1} - x_{n+1}\| \le \|v_{n+1} - \Psi_{n+1}(x_n)\| + \|\Psi_{n+1}(x_n) - x_{n+1}\|$$

$$\le \left(\frac{n}{n+1}\right)^a \|v_n - x_n\| + \frac{1}{(n+1)^{1+b}}.$$

Let $d_n = \|v_n - x_n\|$. One thus has:

$$d_{n+1}(n+1)^a \le d_n n^a + \frac{1}{(n+1)^{(1+b-a)}}.$$

If $a > b$, one obtains for some constants $A, A', A"$:

$$d_{n+1}(n+1)^a \le A' + A" \int_N^{n+1} \frac{dz}{z^{(1+b-a)}} \le \frac{A}{(n+1)^{(b-a)}}.$$

Similarly for $a < b$, there exists A with:

$$d_{n+1}(n+1)^a \le d_N N^a + \int_N^{n+1} \frac{dz}{z^{(1+b-a)}} \le A$$

and for $a = b$:

$$d_{n+1}(n+1)^a \le A' + A" \int_N^{n+1} \frac{dz}{z} \le ALog(n).$$

∎

C.3.2 Discounted version

Consider a family of operators Ψ_λ, $\lambda > 0$, on a Banach space X having the following contracting property:

$$\|\Psi_\lambda(x) - \Psi_\lambda(x')\| \le (1-\lambda)^a \|x - x'\|$$

for some positive a, and λ small enough.
For example $\mathbf{\Psi}$ is non-expansive and $\Psi_\lambda(x) = \lambda \mathbf{\Psi}(\frac{1-\lambda}{\lambda} x)$.
Let v_λ be the fixed point of Ψ_λ.

Theorem C.10
If a family x_λ satisfies, for some positive b, and λ small enough:

$$\|\Psi_\lambda(x_\lambda) - x_\lambda\| \le \lambda^{1+b}$$

and converges strongly to x, as λ goes to 0, then v_λ converges strongly to x also.
Moreover, for any $0 < b' < b$ there exists a constant A such that:

$$\|v_\lambda - x_\lambda\| \le A\lambda^{b'}.$$

Proof
Write, for λ small enough:

$$\|v_\lambda - x_\lambda\| \le \|v_\lambda - \Psi_\lambda(x_\lambda)\| + \|\Psi_\lambda(x_\lambda) - x_\lambda\|$$
$$\le (1-\lambda)^a \|v_\lambda - x_\lambda\| + \lambda^{1+b}.$$

Hence:

$$\|v_\lambda - x_\lambda\| \le \frac{\lambda^{1+b}}{1-(1-\lambda)^a}$$

and the result follows.

■

C.4 Repeated games: Recursive structure and operators

C.4.1 Recursive formula

Consider a two person zero sum repeated game Γ on a state space Ω with current payoff g from $X \times Y \times \Omega$ to $[0,1]$ and transition q from $X \times Y \times \Omega$ to $\Delta(\Omega)$. At each stage, given the state Ω, player 1 (resp. 2) plays in X (resp. Y), the stage payoff is $g(x,y,\omega)$, the new stage ω' is selected according to the distribution $q(x,y,\omega)$ and announced to the players.
Let $\mathcal{F}$ be the set of uniformly bounded functions on Ω. To each $f \in \mathcal{F}$ and each α in $[0,1]$, one associates a game $\Gamma(\alpha, f)$ with strategy spaces X and Y and payoff function in state ω:

$$\mathbf{\Phi}_{xy}(\alpha, f)(\omega) = \alpha g(x,y,\omega) + (1-\alpha) E_{q(x,y,\omega)}(f)$$

Assume that this game has a value on a subset $\mathcal{F}_\Gamma$ of $\mathcal{F}$. One introduces a one parameter family of operators $\mathbf{\Phi}(\alpha, .)$ on $\mathcal{F}_\Gamma$ by :

$$\mathbf{\Phi}(\alpha, f)(\omega) = \mathtt{val}_{X \times Y} \mathbf{\Phi}_{xy}(\alpha, f)(\omega)$$

and another operator by:

$$\mathbf{\Psi}(f)(\omega) = \mathtt{val}_{X \times Y}(g(x,y,\omega) + E_{q(x,y,\omega)}(f))$$

The game $\Gamma(\alpha, f)$ is the one shot game where the current payoff has weight α and the function f evaluates the future. $\mathbf{\Psi}$ is the basic non-expansive **Shapley operator** from which one derives $\mathbf{\Phi}(\alpha, .)$.

Lemma C.11
a) For any constant a:

$$\mathbf{\Phi}(\alpha, f + a) = \mathbf{\Phi}(\alpha, f) + (1-\alpha)a$$

$$\Psi(f+a) = \Psi(f) + a$$

b) Φ and Ψ are monotonic w.r.t. f.
c) $|\Phi(\lambda, f) - \Phi(\mu, f)| \leq |\lambda - \mu| \max(\|f\|_\infty, 1)$

Corollary C.12
a) On the set $(\mathcal{F}_\Gamma, \| \ \|_\infty)$, Φ is contracting with coefficient $(1-\alpha)$ and Ψ is non-expansive.
b) Φ is jointly continuous on $[0,1]\times\mathcal{F}_\Gamma$.

Assume that the n stage game, Γ_n and the game with discount factor λ, Γ_λ have values, respectively denoted by v_n and v_λ.

Properties C.13
a) For $\alpha > 0$,

$$\Phi(\alpha, f) = \alpha\Psi(\frac{(1-\alpha)}{\alpha} f)$$

b)

$$v_n = \Phi(\frac{1}{n}, v_{n-1})$$

b')

$$n v_n = \Psi^n(0)$$

c)

$$v_\lambda = \Phi(\lambda, v_\lambda)$$

c')

$$v_\lambda/\lambda = \Psi((1-\lambda)v_\lambda/\lambda)$$

d) If $(\mathcal{F}_\Gamma, \| \ \|_\infty)$ is complete

$$v_\lambda = (\Phi(\lambda, .))^\infty(f) \text{ for any } f \text{ in } \mathcal{F}_\Gamma$$

Examples
I. Finite stochastic games. (Chapter 5)
The move sets, say I and J, are finite as well as the state space Ω; a transition probability $q(i,j,\cdot)$ from Ω to $\Delta(\Omega)$ and a payoff function g on $I\times J\times\Omega$ are given. Let $X = \Delta(I)$, $Y = \Delta(J)$. Then one has:

$$v_\lambda(\omega) = \mathtt{val}_{X\times Y}\{\lambda \sum\nolimits_{ij} x_i y_j g(i,j,\omega)$$

$$+(1-\lambda)\sum\nolimits_{ij} x_i y_j q(i,j;w)(\omega')v_\lambda(\omega')\}$$

and similarly for v_n.
(Note that we did not specify the signalling structure, we only assumed ω to be known at each stage by both players).

II. Games with incomplete information and standard signalling. (Chapter 4) I and J are the finite move sets. K and L are the finite type sets for Player 1 and Player 2; p, q are the corresponding initial distributions. The payoff matrix in state k, l is A^{kl}. Then:

$$v_\lambda(p,q) = \mathtt{val}_{s\in\Delta(I)^K \times t\in\Delta(J)^L}\{\lambda\sum\nolimits_{ijkl} p^k q^l s_i^k A_{ij}^{kl} t_j^l$$

$$+(1-\lambda)\sum\nolimits_{ij} \overline{s}_i \overline{t}_j v_\lambda(p(i), q(j))\}$$

where $\overline{s} = \sum_k p^k s^k$ and $p(i)$ is the conditional probability given i (and similarly for t and $q(j)$).
Here $X = \Delta(I)^K$, $Y = \Delta(J)^L$, $\Omega = \Delta(K)\times\Delta(L)$ and $q(x, y, (p, q))$ gives probability $\overline{s}_i\overline{t}_j$ to the new state $(p(i), q(j))$.
A similar formula holds for v_n.
III. Finite public case.
This corresponds to a finite stochastic game (I and J are the finite pure move sets) on a finite parameter space Z, with lack of information on both sides; K being the set of types of Player 1 and L of Player 2, p and q being the initial distributions. Assume move/parameter/type dependent payoff A and transition π on Z. In addition the players get signals and we assume that the random signal of each player contains the new parameter (and his own move) and allows to compute the posterior probability of the opponent on his unknown state variable.

$$v_\lambda(p,q,z) = \mathtt{val}_{s\in\Delta(I)^K \times t\in\Delta(J)^L}\{\lambda\sum\nolimits_{ijkl} p^k q^l s_i^k A_{ij}^{klz} t_j^l$$

$$+(1-\lambda)E_{p,q,s,t,z} v_\lambda(p(\alpha,\beta), q(\alpha,\beta), z(\alpha,\beta))\}$$

where α and β are the signals to the players, $z(\alpha,\beta)$ the corresponding new parameter in Z and $p(\alpha,\beta)$ and $q(\alpha,\beta)$ the new conditional distributions on K and L given the signals. The public hypothesis means that, for any (α,β) and (α,β') having positive probability under (s,t), the conditional probability on K is the same given β or β' (and is $p(\alpha,\beta)$) and similarly for q.
Here $X = \Delta(I)^K$ and $Y = \Delta(J)^L$, $\Omega = \Delta(K)\times\Delta(L)\times Z$. The transition on Ω is defined by the distribution on the signals.

Remarks

The "public" hypothesis implies that one can keep Ω as "universal belief space" (see MSZ, Chapter III): at each stage, each player can compute the new beliefs of his opponent on his unknown variable. He does not have to introduce private beliefs on those.
The analysis extends to the dependent case and to the case where Z is a standard Borel measurable space, (see MSZ, Chapter VII, p. 389).

C.4.2 Basic properties of Φ

The main property is expressed by the following domination relation:

Proposition C.14

$$\text{If } \Phi(\lambda, f) \leq f + \delta, \text{ then } v_\lambda \leq f + \delta/\lambda.$$

Proof
Let $\rho = \sup_\omega (v_\lambda - f - \delta/\lambda)^+(\omega)$. Then:

$$(v_\lambda - f - \delta/\lambda) \leq \Phi(\lambda, v_\lambda) - \Phi(\lambda, f) + \delta - \delta/\lambda$$

but $v_\lambda \leq f + \delta/\lambda + \rho$. Thus by Lemma C.11 a) and b):

$$\Phi(\lambda, v_\lambda) \leq \Phi(\lambda, f + \delta/\lambda + \rho) \leq \Phi(\lambda, f) + (1-\lambda)\delta/\lambda + (1-\lambda)\rho$$

hence:

$$(v_\lambda - f - \delta/\lambda) \leq (1-\lambda)\rho$$

a contradiction, if $\rho > 0$.

∎

The inequality says that f is, up to δ, superharmonic w.r.t. $\Phi(\lambda, .)$.

Let now the "per stage error" $\delta\varepsilon$ be proportional to the "per stage weight" ε. Define C_δ^+ as the set of functions f for which there exists a positive ε_0 such that:

$$f + \delta\varepsilon \geq \Phi(\varepsilon, f), \ \forall \varepsilon \in (0, \varepsilon_0)$$

and similarly C_δ^- as the set of f, such that $\exists \varepsilon_0$ with :

$$f - \delta\varepsilon \leq \Phi(\varepsilon, f), \ \forall \varepsilon \in (0, \varepsilon_0)$$

$f \in C_\delta^+$ is approximately superharmonic for all maps $\Phi(\varepsilon, .)$ for ε small enough. Such f are δ **superuharmonic** (u is for uniform). Then one obtains:

Corollary C.15
If f belongs to C_δ^+, then:

$$f + \delta \geq \limsup_{\lambda \to 0} v_\lambda.$$

If f belongs to C_δ^-, then:

$$f - \delta \leq \liminf_{\lambda \to 0} v_\lambda.$$

Proof
f belonging to C_δ^+ implies that for all λ small enough $\mathbf{\Phi}(\lambda, f) \leq f + \delta\lambda$, thus by Proposition C.14, $v_\lambda \leq f + \delta$. ∎

A similar result holds for the finite game.

Lemma C.16
If f belongs to C_δ^+, then:

$$f + \delta \geq \limsup_{n\to\infty} v_n.$$

If f belongs to C_δ^-, then:

$$f - \delta \leq \liminf_{n\to\infty} v_n.$$

Proof
Let $N \geq 1/\varepsilon_0$. Let us prove by induction that:

$$v_n \leq f + \frac{n+1-N}{n}\delta + \frac{N-1}{n}\|f - v_{N-1}\|_\infty.$$

In fact by Lemma C. 14:

$$v_N = \mathbf{\Phi}(\frac{1}{N}, v_{N-1}) \leq \mathbf{\Phi}(\frac{1}{N}, f) + \frac{N-1}{N}\|f - v_{N-1}\|_\infty$$

and

$$\mathbf{\Phi}(\frac{1}{N}, f) \leq f + \frac{\delta}{N}.$$

Now at step $n+1$:

$$\begin{aligned} v_{n+1} &= \mathbf{\Phi}(\frac{1}{n+1}, v_n) \\ &\leq \mathbf{\Phi}(\frac{1}{n+1}, f) + \frac{n}{n+1}(\frac{n+1-N}{n}\delta + \frac{N-1}{n}\|f - v_{N-1}\|_\infty) \\ &\leq f + \delta(\frac{n+2-N}{n+1}) + \frac{N-1}{n}\|f - v_{N-1}\|_\infty. \end{aligned}$$

∎

Notations
Let $\mathcal{C}^+ = \cap_{\delta>0}\mathcal{C}_\delta^+$ and similarly $\mathcal{C}^- = \cap_{\delta>0}\mathcal{C}_\delta^-$. Finally define $\mathcal{C} = \overline{\mathcal{C}^+}\cap\overline{\mathcal{C}^-}$ where, for or a subset $\mathcal{G}$ of $\mathcal{F}$, $\overline{\mathcal{G}}$ denotes its closure for the uniform norm. Then one obtains:

Proposition C.17
If $f \in \mathcal{C}$, then:

$$f = \lim_{\lambda \to 0} v_\lambda = \lim_{n \to \infty} v_n.$$

In particular C has at most one element.

Proof
The proof follows from Corollary C.15 and Lemma C.16. ∎

C.4.3 The derived game

Assume from now on:
i) X and Y are compact Hausdorff.
The function $(x, y) \mapsto \mathbf{\Phi}_{xy}(\alpha, f)(\omega)$ is u.s.c. in x and l.s.c. in y.
The game $\Gamma(\alpha, f)(\omega)$ has a value. (This is in particular the case if X and Y are convex and $(x, y) \mapsto \mathbf{\Phi}_{xy}(\alpha, f)(\omega)$ is quasi-concave in x, quasi-convex in y, by Theorem A.7).
Denote by $X(\alpha, f)(\omega)$ and $Y(\alpha, f)(\omega)$ the corresponding sets of optimal strategies.
ii) the function

$$(x, y) \mapsto \varphi_{xy}(f)(\omega) = g(x, y, \omega) - E_{q(x,y,\omega)}(f)$$

is u.s.c. in x, l.s.c. in y on $X \times Y$.

Following Appendix A.7, we associate to the "directional derivative" of the value of the game $\Gamma(\alpha, f)$ a new game as follows:

Definition
The **derived game** $\partial\Gamma(f)(\omega)$ is the game with payoff $\varphi_{xy}(f)(\omega)$ played on $X(0, f)(\omega) \times Y(0, f)(\omega)$.
The main property is the following:

Proposition C.18
$\partial\Gamma(f)(\omega)$ has a value, denoted by $\varphi(f)(\omega)$, and optimal strategies.
Moreover it satisfies:

$$\varphi(f)(\omega) = \lim_{\alpha \to 0^+} \frac{\mathbf{\Phi}(\alpha, f)(\omega) - \mathbf{\Phi}(0, f)(\omega)}{\alpha}.$$

Proof
Recall that $\mathbf{\Phi}_{xy}(0, f)(\omega) = E_{q(x,y,\omega)}(f)$ so that

$$\mathbf{\Phi}_{xy}(\alpha, f)(\omega) = \mathbf{\Phi}_{xy}(0, f)(\omega) + \alpha\varphi_{xy}(f)(\omega).$$

For ω given, apply Proposition A.15 to the game $\Gamma(\alpha, f)(\omega)$. ∎

$\Gamma(0,f)$ appears as the **projective game**: if f represents the current "level", the payoff is the expectation of the level tomorrow: $E_{q(x,y,\omega)}(f)$. $\partial\Gamma(f)$ is the derivative of $\Gamma(\alpha,f)$ at 0. The payoff $\varphi_{xy}(f)$ in the derived game measures the difference between the current payoff g and the expected future "level" when both players play optimally in the projective game.

A useful consequence of the previous Proposition is the following property: if a strategy x is good in the projective game, it cannot guarantee more than the value in the derived game.

Corollary C.19

$\forall\omega, \forall\rho > 0, \exists\eta > 0$ such that, $\forall x \in X$, either:
a) $\exists y \in Y$ with $\mathbf{\Phi}_{xy}(0,f)(\omega) \leq \mathbf{\Phi}(0,f)(\omega) - \eta$
or
b) for any y optimal in $\partial\Gamma(f)(\omega)$: $\varphi_{xy}(f)(\omega) \leq \varphi(f)(\omega) + \rho$.

Proof

The proof is by contradiction. Otherwise for a specific ω and some $\rho > 0$ one could find for each positive integer m, a strategy x_m in X with

$$\mathbf{\Phi}_{x_m y}(0,f)(\omega) > \mathbf{\Phi}(0,f)(\omega) - 1/m, \ \forall y \in Y$$

and

$$\varphi_{x_m y}(f)(\omega) > \varphi(f)(\omega) + \rho, \text{ for some } y_m \text{ optimal in } \partial\Gamma(f)(\omega).$$

If x^* (resp. y^*) is an accumulation point of the sequence x_m (resp. y_m), the first inequality:

$$\mathbf{\Phi}_{x^* y}(0,f)(\omega) \geq \mathbf{\Phi}(0,f)(\omega), \ \forall y \in Y$$

shows that x^* belongs to $X(0,f)(w)$ which together with the second inequality:

$$\varphi_{x^* y^*}(f)(\omega) > \varphi(f)(\omega) + \rho, \text{ for some } y^* \text{ optimal in } \partial\Gamma(f)(\omega)$$

contradicts the definition of $\varphi(f)(\omega)$. ∎

Properties C.13 imply that any uniform limit of v_n or of v_λ will satisfy $\mathbf{\Phi}(0,f) = f$. It is thus natural to consider also the ratio $\dfrac{\mathbf{\Phi}(\alpha,f)(\omega) - f(\omega)}{\alpha}$.

Proposition C.20

$$\varphi^*(f)(w) = \lim_{\alpha\to 0} \frac{\mathbf{\Phi}(\alpha,f)(\omega) - f(\omega)}{\alpha}$$

exists in $\mathbb{R} \cup \{-\infty, +\infty\}$.

Proof
If $\Phi(0, f)(\omega) < f(\omega)$, $\varphi^*(f)(\omega) = -\infty$ and if $\Phi(0, f)(\omega) > f(\omega)$, then $\varphi^*(f)(\omega) = +\infty$.
Finally if $\Phi(0, f)(\omega) = f(\omega)$, $\varphi^*(f)(\omega) = \varphi(f)(\omega)$. ∎

These operators will be useful to define new subsets of functions.

Definition
Let $\mathcal{S}^+$ be the set of functions satisfying the following system:

$$\Phi(0, f) \leq f \qquad \text{and} \qquad \Phi(0, f)(\omega) = f(\omega) \Rightarrow \varphi(f)(\omega) \leq 0$$

or equivalently

$$\varphi^*(f) \leq 0.$$

Let similarly $\mathcal{S}^-$ be the set defined by:

$$\Phi(0, f) \geq f \qquad \text{and} \qquad \Phi(0, f)(\omega) = f(\omega) \Rightarrow \varphi(f)(\omega) \geq 0$$

or equivalently

$$\varphi^*(f) \geq 0 .$$

These new sets relate to the ones introduced in Part B as follows.

Proposition C.21

$$\mathcal{C}^+ \subset \mathcal{S}^+.$$

Proof
Let $f \in \mathcal{C}^+$. Then for any δ there is an $\varepsilon(\delta)$ such that for all $\varepsilon \leq \varepsilon(\delta)$ and all $\omega \in \Omega$,

$$\Phi(\varepsilon, f)(\omega) \leq f(\omega) + \delta\varepsilon.$$

By letting ε go to 0, one gets

$$\varphi^*(f)(\omega) \leq \delta.$$

∎

In fact the reverse inclusion holds in the finite case.

Proposition C. 22
Assume Ω is finite.

$$\mathcal{S}^+ \subset \mathcal{C}^+ \text{and} \mathcal{S}^- \subset \mathcal{C}^-.$$

Proof
Let $f \in \mathcal{S}^+$. Fix $\omega \in \Omega$. Hence $\varphi^*(f)(\omega) \leq 0$, so that by Proposition C.20 for any $\delta > 0$ there exists $\varepsilon_0(\omega) > 0$ such that $\varepsilon \leq \varepsilon_0(\omega)$ implies:

$$\Phi(\varepsilon, f)(\omega) - f(\omega) \leq \delta\varepsilon.$$

Since Ω is finite the conclusion follows. ∎

We obtain in this case a "local" condition implying that f belongs to $\mathcal{C}$, namely:

Corollary C.23
Assume Ω finite. If f is in the closure of both $\mathcal{S}^+$ and $\mathcal{S}^-$, then f belongs to $\mathcal{C}$. In particular $\overline{\mathcal{S}}^+ \cap \overline{\mathcal{S}}^-$ contains at most one point.

In the remaining of this section, we provide conditions extending this result to a compact state space Ω.
The next proposition compares the variation $\Phi(\alpha, f) - f$ for two functions at a point ω maximizing their difference. A similar property was proved by Kohlberg (1974) for the case of constant functions.

Proposition C.24 (Maximal principle)
i) Let f_1 and f_2 and ω satisfy:

$$f_2(\omega) - f_1(\omega) = \delta = \max_{\omega' \in \Omega} (f_2 - f_1)(\omega') > 0$$

then:

$$(\Phi(\alpha, f_1)(\omega) - f_1(\omega)) - (\Phi(\alpha, f_2)(\omega) - f_2(\omega)) \geq \alpha(f_2(\omega) - f_1(\omega))$$

hence:

$$\varphi^*(f_1)(\omega) - \varphi^*(f_2)(\omega) \geq \delta.$$

ii) If moreover:

$$\Phi(0, f_2)(\omega) \geq f_2(\omega), \Phi(0, f_1)(\omega) \leq f_1(\omega)$$

then:

$$\Phi(0, f_i)(\omega) = f_i(\omega), i = 1, 2$$

so that:

$$\varphi(f_1)(\omega) - \varphi(f_2)(\omega) \geq \delta.$$

Proof
For any $\omega' \in \Omega$:

$$\begin{aligned}\Phi(\alpha, f_2)(\omega') - \Phi(\alpha, f_1)(\omega') &\leq \Phi(\alpha, f_1 + \delta)(\omega') - \Phi(\alpha, f_1)(\omega') \\ &\leq (1 - \alpha)\delta \\ &\leq (1 - \alpha)(f_2(\omega) - f_1(\omega)).\end{aligned}$$

So that in particular:

$$(\Phi(\alpha, f_1)(\omega) - f_1(\omega)) - (\Phi(\alpha, f_2)(\omega) - f_2(\omega)) \geq \alpha(f_2(\omega) - f_1(\omega)).$$

Hence dividing by α and letting α go to 0, using Proposition C.20 one has :

$$\varphi^*(f_1)(\omega) - \varphi^*(f_2)(\omega) \geq \delta.$$

For $ii)$, taking $\alpha = 0$ in the previous inequality implies $\mathbf{\Phi}(0, f_i)(\omega) = f_i(\omega), i = 1, 2$ so that one obtains as well:

$$(\mathbf{\Phi}(\alpha, f_2)(\omega) - \mathbf{\Phi}(0, f_2)(\omega)) - (\mathbf{\Phi}(\alpha, f_1)(\omega) - \mathbf{\Phi}(0, f_1)(\omega)) \geq \alpha(f_2(\omega) - f_1(\omega))$$

hence :

$$\varphi(f_1)(\omega) - \varphi(f_2)(\omega) \geq \delta.$$

∎

This result allows to compare functions in $\mathcal{S}^+$ and $\mathcal{S}^-$ in the continuous case.

Proposition C.25
Assume Ω compact. For all continuous functions $f_1 \in \mathcal{S}^+$ and $f_2 \in \mathcal{S}^-$ one has:

$$f_1(\omega) \geq f_2(\omega), \quad \forall \omega \in \Omega.$$

Proof
Otherwise, let ω be a point realizing the maximum of $(f_2 - f_1)$ on Ω and assume $(f_2 - f_1)(\omega) = \delta > 0$. By Proposition C.24, $\varphi^*(f_1)(\omega) - \varphi^*(f_2)(\omega) \geq \delta$, hence a contradiction to $\varphi^*(f_1) \leq 0$ and $\varphi^*(f_2) \geq 0$ on Ω. ∎

Hence we obtain also uniqueness in this case.

Corollary C.26
Assume Ω compact. Let $\mathcal{S}_0^+$ (resp. $\mathcal{S}_0^-$) be the subset of continuous functions on Ω belonging to $\mathcal{S}^+$ (resp. $\mathcal{S}^-$).
The closures of $\mathcal{S}_0^+$ and $\mathcal{S}_0^-$ have at most one common element.

In the same spirit as in Proposition C.24 one has:

Proposition C.27 (Tightness)
Let f_1, f_2 and ω satisfy:

$$f_2(\omega) - f_1(\omega) = \delta = \max_{w' \in \Omega} (f_2 - f_1)(\omega') > 0$$

and

$$\mathbf{\Phi}(0, f_2)(\omega) \geq f_2(\omega), \quad \mathbf{\Phi}(0, f_1)(\omega) \leq f_1(\omega).$$

Then, for any $\theta \in [0, 1]$:

$$\mathbf{\Phi}(0, \theta f_2 + (1 - \theta) f_1)(\omega) = \theta f_2(\omega) + (1 - \theta) f_1(\omega)$$

and

$$X(0, f_2)(\omega) \subset X(0, \theta f_2 + (1 - \theta) f_1)(\omega)$$

$$Y(0, f_1)(\omega) \subset Y(0, \theta f_2 + (1-\theta) f_1)(\omega).$$

Proof
One has:

$$\mathbf{\Phi}(0, \theta f_2 + (1-\theta) f_1) \;\leq \mathbf{\Phi}(0, f_1 + \theta\delta) \;\leq\; f_1 + \theta\delta$$

hence at ω:

$$\mathbf{\Phi}(0, \theta f_2 + (1-\theta) f_1)(\omega) \;\leq\; \theta f_2(\omega) + (1-\theta) f_1(\omega)$$

and a dual inequality holds.
Let $x \in X(0, f_2)(\omega)$. Then:

$$\begin{aligned}
\mathbf{\Phi}_{xy}(0, \theta f_2 + (1-\theta) f_1)(w) &\geq \mathbf{\Phi}_{xy}(0, f_2 - (1-\theta)\delta)(\omega) \\
&\geq \mathbf{\Phi}_{xy}(0, f_2)(\omega) - (1-\theta)\delta \\
&\geq f_2(\omega) - (1-\theta)\delta \\
&\geq \theta f_2(\omega) + (1-\theta) f_1(\omega) \\
&= \mathbf{\Phi}(0, \theta f_2 + (1-\theta) f_1)(w)
\end{aligned}$$

hence the result.

∎

C.4.4 Comments

See MSZ, Chapter IV for the general model of recursive structure and specific formulations.
Consult also MSZ (pages 207-212; 298-301; 387-389; 408-409; 418-422) and the references therein for similar constructions and properties of operators.

The derived game and the corresponding operator were introduced in the study of absorbing games (where Ω is reduced to one point) by Kohlberg (1974), see Chapter 5, Section 5.5. Most of results here follow Rosenberg and Sorin (2001).

D: Kuhn's Theorem for repeated games

Mixed and behavioral strategies in repeated games are introduced and a simple version of Kuhn's theorem is provided.

A general model of **repeated game** is defined as follows. (We first consider the finite case where all sets to be introduced are finite.)
The **state** space is Ω, the **signal**'s space for each player i, $i \in I$, is L^i and his set of **moves** is A^i. Let $L = \prod_i L^i$ and $A = \prod_i A^i$. A probability π on $\Omega \times L$, a transition probability q from $\Omega \times A$ to $\Omega \times L$ and a payoff function g from $\Omega \times A$ to $\mathbb{R}^{\# I}$ are given. The game is played in stages.
At stage 0, a vector (ω_1, ℓ_1) is chosen according to $\pi \in \Delta(\Omega \times L)$. ω_1 is the initial state. Each player i is told his signal ℓ_1^i which is the ith component of ℓ_1.
At stage one, each player i chooses a move a_1^i in A^i and a new vector (w_2, ℓ_2) is selected according to $q(\omega_1, a_1) \in \Delta(\Omega \times L)$. The payoff is $g_1 = g(\omega_1, a_1)$. Again each player i is informed only upon ℓ_2^i.
Similarly at stage n, each player i chooses a move a_n^i in A^i the stage payoff is $g_n = g(\omega_n, a_n)$, (w_{n+1}, ℓ_{n+1}) is selected according to $q(\omega_n, a_n)$, ω_{n+1} is the new state and the vector of signals is ℓ_{n+1}.

The natural **tree structure** of the game is given by the sets of finite **histories** $H = \cup_{n \geq 1} H_n$ with $H_n = (\Omega \times L \times A)^{n-1} \times \Omega \times L$ being the set of histories before the n-th move. Let also $H_\infty = (\Omega \times L \times A)^\infty$ be the set of **plays**.

Let us give few examples of such games:
Stochastic games correspond to the case where for each i, $\ell_1^i = \omega_1$ and $\ell_{n+1}^i = \{a_n, \omega_{n+1}\}, n \geq 1$.
Stochastic games with signals deal with the case where for each i, ℓ_n^i reveals ω_n: the state is known.
Incomplete information games in the standard case are defined by a constant state ($\omega_1 = \omega_n$) and for each i, $\ell_{n+1}^i = a_n$.
Incomplete information games with signals (level 3, see Chapter 3, 7.3) correspond to a signalling structure where the law of ℓ_{n+1} is $q(\omega_1, a_n)$.

In defining the strategies of the players we will assume that the game is with **perfect recall**. In the current framework this means that each player recalls the signals he got and the moves he played. We are thus led to define the **histories for player** i as $H^i = \cup_{n\geq 1} H^i_n$ with $H^i_n = (L^i \times A^i)^{n-1} \times L^i$ describing his **information** prior to his n-th move, $h^i_n = (\ell^i_1, a^i_1, ..., \ell^i_n)$.

We introduce now three sets of functions:
A **pure strategy** for player i is a map θ^i from H^i to A^i, or equivalently a collection of maps θ^i_n from H^i_n to A^i. $\theta^i(h^i_n)$ is the n-th move of player i after learning history h^i_n.
A **behavioral strategy** for player i is a map σ^i from H^i to $\Delta(A^i)$, or equivalently a collection of maps σ^i_n from H^i_n to $\Delta(A^i)$. $\sigma^i(h^i_n)$ is the distribution of the n-th move of player i after knowing history h^i_n, $\sigma^i(h^i_n)[a^i]$ being the probability of playing move a^i.
A **mixed strategy** for player i is a distribution ζ^i on the set Θ^i of pure strategies of player i.

Note that a I-vector of strategies α, where each α^i is a pure or a behavioral strategy, induces with π and q a unique probability on H. Explicitly if $h = (\omega_1, \ell_1, a_1, ..., a_{n-1}, \omega_n, \ell_n)$:

$$\mathbb{P}_\alpha(h) = \pi(\omega_1, \ell_1) \prod_i \alpha^i(h^i_1)[a^i_1] q(\omega_1, \ell_1)[\omega_2, \ell_2] \prod_i \alpha^i(h^i_2)[a^i_2] ...$$

$$...q(\omega_{n-1}, a_{n-1})[\omega_n, \ell_n]$$

where $\alpha^i(h^i_n)[a^i_n]$ is either $\mathbf{1}_{\{\theta^i(h^i_n)=a^i_n\}}$ or $\sigma^i(h^i_n)[a^i_n]$. This probability is consistent and has a unique extension $\mathbb{P}_\alpha$ to $(H_\infty, \mathcal{H}_\infty)$ where the σ-algebra $\mathcal{H}_\infty$ is generated by the cylinders $h_n \times H_\infty$.
Similarly the measurable structure on Θ^i, needed to define the mixed strategies, is generated by the sets $\theta^i[n]$ of pure strategies that coincide with some strategy θ^i up to some stage n. Hence for a vector α of any kind of strategies the corresponding distribution on plays $\mathbb{P}_\alpha$ is well defined.
The next fundamental result is due to Kuhn (1953).

Theorem D.1
Given any vector $\alpha = \{\alpha^i\}_{i\in I}$ where each α^i is a pure or behavioral or mixed strategy, for each i there exists a behavioral strategy σ^i and a mixed strategy ζ^i such that:

$$\mathbb{P}_\alpha = \mathbb{P}_{(\sigma^i, \alpha^{-i})} = \mathbb{P}_{(\zeta^i, \alpha^{-i})}$$

In words, whatever being the behavior of the opponents, given a behavioral (resp. mixed) strategy, there exists a mixed (resp. behavioral) strategy which is equivalent in the sense that it induces the same distribution on plays.

Proof
Take $i = 1$. Starting from a behavioral strategy σ^1, it is enough, in order to specify ζ^1, to define $\zeta^1(\theta^1[n])$, for each θ^1 and each n. We simply compute

the probability that σ^1 coincides with θ^1. Note that, by definition of the pure strategies:

$$\zeta^1(\theta^1[n]) = \prod_{h^1 \in H_n^1} \sigma^1(h_1^1)[\theta^1(h_1^1)]\sigma^1(h_2^1)[\theta^1(h_2^1)]....\sigma^1(h_n^1)[\theta^1(h_n^1)]$$

defines a probability as θ^1 varies. It is then clear that $\mathbb{P}_{(\sigma^1,\alpha^{-1})} = \mathbb{P}_{(\zeta^1,\alpha^{-1})}$ on H_n.
On the other hand, to represent a mixed strategy by a behavioral one we first introduce, given h_n^1 the subset $\Theta^1(h_n^1)$ of pure strategies compatible with h_n^1. Namely if $h_n^1 = (\ell_1^1, a_1^1, \ell_2^1, a_2^1, ..., a_{n-1}^1, \ell_n^1)$, then

$$\Theta^1(h_n^1) = \{\theta^1 : \theta^1(\ell_1^1, a_1^1, \ell_2^1, a_2^1, ..., a_{m-1}^1, \ell_m^1) = a_m^1, m = 1, ..., n\}$$

The probability of choosing then move a^1 is defined, if $\Theta^1(h_n^1) \neq \emptyset$ by:

$$\sigma^1(h_n^1)[a^1] = \zeta^1(\{\theta^1; \theta^1(h_n^1) = a^1\} | \Theta^1(h_n^1))$$

and is arbitrarily otherwise.Then on each history for player 1 the product of conditional probabilities will generate the right probability as induced by ζ^1.

■

Let us now consider the general case where $(\Omega, \mathcal{C})$, $(L^i, \mathcal{L}^i)$, $(A^i, \mathcal{A}^i)$ are measurable spaces (Standard Borel) and H^i is endowed with the product σ-field $\mathcal{H}^i$.
θ^i is then a measurable map from $(H^i, \mathcal{H}^i)$ to $(A^i, \mathcal{A}^i)$.
Similarly σ^i is a transition probability from $(H^i, \mathcal{H}^i)$ to $(A^i, \mathcal{A}^i)$.
To define ζ^i we need a measurable structure on Θ^i. To avoid this problem we follow Aumann (1964) in introducing an auxiliary space $(X^i, \mathcal{X}^i)$, which can be taken as a copy of $[0, 1]$ with the Borel σ-algebra $\mathcal{B}$ and Lebesgue measure Λ and we represent also $(A^i, \mathcal{A}^i)$ as $([0, 1], \mathcal{B})$.
ζ^i is now a measurable map from $(H^i \times X^i, \mathcal{H}^i \otimes \mathcal{X}^i)$ to $(A^i, \mathcal{A}^i)$. The interpretation is that x is chosen at random according to Λ and then the pure strategy $\zeta^i(., x)$ is played.
The reduction from behavioral strategies to mixed strategies works as follows. Given σ^i, $h \in H^i$ and $a \in A^i$ let $Q(\sigma^i, h)(a) = \sigma^i(h)([0, a])$ be the cumulative function of the distribution $\sigma^i(h)$. Define ζ^i by:

$$\zeta^i(h, x) = \min\{a, Q(\sigma^i, h)(a) \geq x\}$$

then clearly $\sigma^i(h)$ and $\zeta^i(h, .)$ induce the same distribution on A^i. To get the same result for each h one replaces X^i by a countable product X_n^i of i.i.d. random variables x_n being used at stage n.
For the other direction the difficulty lies in the fact that the above sets $\Theta^i(h_n^i)$ have measure zero and one deals with regular conditional probabilities. We refer to Aumann (1964) and MSZ (Chapter II, 1.c).

Comments

One could define even more general strategies, namely mixtures of behavioral strategies (MSZ, II, 1.c): μ^i is a probability on Σ^i.
There are in fact two parts in Kuhn's Theorem:
The representation of a general strategy as a mixed one, that uses only the fact that the game is **linear** (Isbell, 1957): each information set is crossed at most once on any play.
The representation of mixed strategies by behavioral ones under the perfect recall assumption.
One can also define the perfect recall property for each player individually and then obtain the equivalence property for this player.

In the next examples game A is not linear while games B and C do not respect perfect recall.

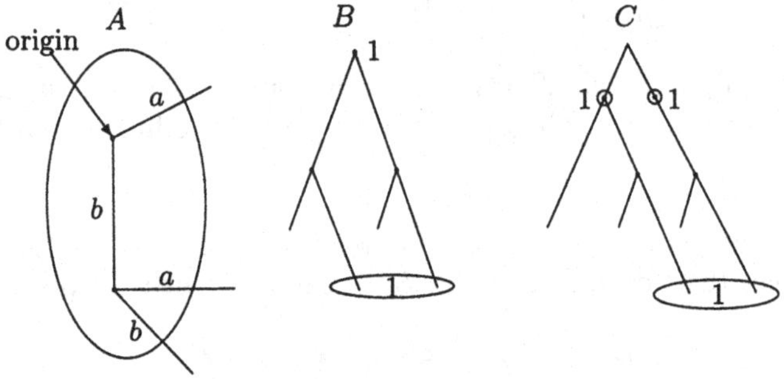

Fig. D.1.

In game B the player forgets his initial move while in game C he does not remember his previous information.

References

Aumann R.J. (1964) Mixed and behavior strategies in infinite extensive games, in *Advances in Game Theory*, M. Dresher, L.S. Shapley and A. W. Tucker (eds.), Annals of Mathematical Studies 52, Princeton University Press 627-650.

Aumann R.J. and M. Maschler (1995) (with the collaboration of R. Stearns) *Repeated Games with Incomplete Information*, M.I.T. Press.

Benedetti R. and J.-J. Risler (1990) *Real Algebraic and Semi-Algebraic Sets*, Hermann.

Berge C. (1966) *Espaces Topologiques, Fonctions Multivoques*, Dunod.

Bergin J. (1992) Player type distributions as state variables and information revelation in zero sum repeated games with discounting, *Mathematics of Operations Research*, **17**, 640-656.

Bewley T. and E. Kohlberg (1976a) The asymptotic theory of stochastic games, *Mathematics of Operations Research*, **1**, 197-208.

Bewley T. and E. Kohlberg (1976b) The asymptotic solution of a recursion equation occurring in stochastic games, *Mathematics of Operations Research*, **1**, 321-336.

Bewley T. and E. Kohlberg (1978) On stochastic games with stationary optimal strategies *Mathematics of Operations Research* **3**, 104-125.

Blackwell D. (1956) An analog of the minmax theorem for vector payoffs, *Pacific Journal of Mathematics*, **6**, 1-8.

Blackwell D. (1962) Discrete dynamic programming, *Annals of Mathematical Statistics*, **33**, 719-726.

Blackwell D. and T. Ferguson (1968) The "Big Match", *Annals of Mathematical Statistics*, **39**, 159-163.

Borel E. (1921) La théorie du jeu et les équations intégrales à noyau symétrique gauche, *C.R.A.S.*, **173**, 1304-1308.

Coulomb J.-M. (1992) Repeated games with absorbing states and no signals, *International Journal of Game Theory*, **21**, 161-174.

Coulomb J.-M. (1996) A note on 'Big Match', *ESAIM: Probability and Statistics*, **1**, 89-93.

Coulomb J.-M. (1999) Generalized Big Match, *Mathematics of Operations Research*, **24**, 795-816.

Coulomb J.-M. (2001) Absorbing games with a signalling structure, *Mathematics of Operations Research*, **26**, 286-303.

Couwenberg H.A.M. Stochastic games with metric spaces, *International Journal of Game Theory*, **9**, 25-36.

De Meyer B. (1996a) Repeated games and partial differential equations, *Mathematics of Operations Research*, **21**, 209-236.

De Meyer B. (1996b) Repeated games, duality and the Central Limit theorem, *Mathematics of Operations Research*, **21**, 237-251.

De Meyer B. (1998) The maximal variation of a bouded martingale and the Central Limit theorem, *Annales de l'Institut Henri Poincaré, Probabilités et Statistiques*, **34**, 49-59.

De Meyer B. (1999) ¿From repeated games to Brownian games, *Annales de l'Institut Henri Poincaré, Probabilités et Statistiques*, **35**, 1-48.

De Meyer B. and D. Rosenberg (1999) "Cav u" and the dual game, *Mathematics of Operations Research*, **24**, 619-626.

Dynkin E.B. and A.A. Yushkevich (1979) *Controlled Markov Processes*, Springer.

Everett H. (1957) Recursive games, in *Contributions to the Theory of Games, III*, M. Dresher, A.W. Tucker and P. Wolfe (eds.), Annals of Mathematical Studies, 39, Princeton University Press, 47-78.

Fan K. (1953) Minimax Theorems, *Proceedings of the National Academy of Sciences of the U.S.A*, **39**, 42-47.

Ferguson T. S., L. S. Shapley, and R. Weber (1970) A stochastic game with incomplete information, preprint.

Filar J. and K. Vrieze (1997) *Competitive Markov Decision Processes*, Springer.

Forges F. (1982) Infinitely repeated games of incomplete information symmetric case with random signals, *International Journal of Game Theory*, **11**, 203-213.

Foster 0. (1981) *Lectures on Riemann Surfaces*, Springer.

Frid E.B. (1973) On stochastic games *Theory of Probability and Applications* **18**, 389-393.

Geitner J. (1999) Equilibria in stochastic games of incomplete information: the general symmetric case, preprint.

Gillette D. (1957) Stochastic games with zero stop probabilities, in *Contributions to the Theory of Games, III*, M. Dresher, A.W. Tucker and P. Wolfe (eds.), Annals of Mathematical Studies, 39, Princeton University Press, 179-187.

Heuer M. (1992a) Asymptotically optimal strategies in repeated games with incomplete information, *International Journal of Game Theory*, **20**, 377-392.

Heuer M. (1992b) Optimal strategies for the uninformed player, *International Journal of Game Theory*, **20**, 33-51.

Isbell J.R. (1957) Finitary games, in *Contributions to the Theory of Games, III*, M. Dresher, A.W. Tucker and P. Wolfe (eds.), Annals of Mathematical Studies, 39, Princeton University Press, 79-96.

Kohlberg E. (1974) Repeated games with absorbing states, *Annals of Statistics*, **2**, 724-738.

Kohlberg E. (1975a) The information revealed in infinitely-repeated games of incomplete information, *International Journal of Game Theory*, **4**, 57-59.

Kohlberg E. (1975b) Optimal strategies in repeated games with incomplete information, *International Journal of Game Theory*, **4**, 7-24.

Kohlberg E. and A. Neyman (1981) Asymptotic behavior of non expansive mappings in normed linear spaces, *Israel Journal of Mathematics*, **38**, 269-275.

Kohlberg E. and S. Zamir (1974) Repeated games of incomplete information: the symmetric case, *Annals of Statistics*, **2**,1040-1041.

Krausz A. and U. Rieder (1997) Markov games with incomplete information, *Mathematical Methods of Operations Research*, **46**, 263-279.

Kuhn H. W. (1953) Extensive games and the problem of information, in *Contributions to the Theory of Games, II*, H.W. Kuhn and A.W. Tucker (eds.), Annals of Mathematical Studies, 28, Princeton University Press, 193-216.

Laraki R. (2000a) Repeated games with lack of information on one side: the dual differential approach, to appear in *Mathematics of Operations Research.*

Laraki R. (2000b) The splitting games and applications, to appear in *International Journal of Game Theory.*

Laraki R. (2000c) Duality and games with incomplete information, preprint.

Laraki R. (2001) Variational inequalities, system of functional equations and and incomplete information repeated games, *SIAM Journal on Control and Optimization*, **40**, 516-524.

Lehrer E. (1987) A note on the monotonicity of v_n, *Economic Letters*, **23**, 341-342.

Lehrer E. (1997) Approachability in infinite dimensional spaces and an application: a universal algorithm for generating extended normal numbers, preprint.

Lehrer E. and D. Monderer (1994) Discounting versus averaging in dynamic programming, *Games and Economic Behavior*, **6**, 97-113.

Lehrer E. and S. Sorin (1992) A uniform Tauberian theorem in dynamic programming, *Mathematics of Operations Research*, **17**, 303-307.

Lehrer E. and L. Yariv (1999) Repeated games with incomplete information on one side: the case of different discount factors, *Mathematics of Operations Research*, **24**, 204-218.

Loomis L. H. (1946) On a theorem of von Neumann, *Proceeding of the National Academy of Sciences of the U.S.A*, **32**, 213-215.

Maitra A. and T. Parthasarathy (1970) On stochastic games, *Journal of Optimization Theory and Applications*, **5**, 289-300.

Maitra A. and W. Sudderth (1992) An operator solution of stochastic games, *Israel Journal of Mathematics*, **78**, 33-49.

Maitra A. and W. Sudderth (1993) Borel stochastic games with lim sup payoff, *Annals of Probability*, **21**, 861-885.

Maitra A. and W. Sudderth (1996) *Discrete Gambling and Stochastic Games*, Springer.

Maitra A. and W. Sudderth (1998) Finitely additive stochastic games with Borel measurable payoffs, *International Journal of Game Theory*, **27**, 257-267.

Mayberry J. P. (1967) Discounted repeated games with incomplete information, *Report of the U.S. Arms Control and Disarmament Agency, ST 116, Chapter V*, Mathematica, Princeton, 435-461.

Megiddo N. (1980) On repeated games with incomplete information played by non-Bayesian players, *International Journal of Game Theory*, **9**, 157-167.

Melolidakis C. (1989) On stochastic games with lack of information on one side, *International Journal of Game Theory*, **18**, 1-29.

Melolidakis C. (1991) Stochastic games with lack of information on one side and positive stop probabilities, in *Stochastic Games and Related Topics*, T.E.S. Raghavan and al. (eds.), Kluwer, 113-126.

Mertens J.-F. (1972) The value of two-person zero-sum repeated games: the extensive case, *International Journal of Game Theory*, **1**, 217-227.

Mertens J.-F. (1973) A note on "The value of two-person zero-sum repeated games: the extensive case", *International Journal of Game Theory*, **2**, 231-234.

Mertens J.-F. (1982) Repeated games: an overview of the zero-sum case, in *Advances in Economic Theory*, Hildenbrand W. (ed.), Cambridge University Press, 175-182.

Mertens J.-F. (1986) The minmax theorem for u.s.c.-l.s.c. payoff functions, *International Journal of Game Theory*, **15**, 237-250.

Mertens J.-F. (1987) Repeated games, in *Proceedings of the International Congress of Mathematicians, Berkeley, 1986*, Gleason A. M. (ed.), American Mathematical Society, 1528-1577.

Mertens J.-F. (1998) The speed of convergence in repeated games with incomplete information on one side, *International Journal of Game Theory*, **27**, 343-359.

Mertens J.-F. (2001) Stochastic games, in *Handbook of Game Theory, III*, Aumann R.J. and S. Hart (eds.), North-Holland.

Mertens J.-F. and A. Neyman (1981) Stochastic games, *International Journal of Game Theory*, **10**, 53-66.

Mertens J.-F., S. Sorin and S. Zamir (1994) *Repeated Games*, CORE D.P. 9420-21-22.

Mertens J.-F. and S. Zamir (1971) The value of two-person zero-sum repeated games with lack of information on both sides, *International Journal of Game Theory*, **1**, 39-64.

Mertens J.-F. and S. Zamir (1976a) The Normal distribution and repeated games, *International Journal of Game Theory*, **5**, 39-64.

Mertens J.-F. and S. Zamir (1976b) On a repeated game without a recursive structure, *International Journal of Game Theory*, **5**, 173-182.

Mertens J.-F. and S. Zamir (1977a) A duality theorem on a pair of simultaneous functional equation, *Journal of Mathematical Analysis and Applications*, **60**, 550-558.

Mertens J.-F. and S. Zamir (1977b) The maximal variation of a bounded martingale, *Israel Journal of Mathematics*, **27**, 252-276.

Mertens J.-F. and S. Zamir (1980) Minmax and maxmin of repeated games with incomplete information, *International Journal of Game Theory*, **9**, 201-215.

Mertens J.-F. and S. Zamir (1981) Incomplete information games with transcendental values, *Mathematics of Operations Research*, **6**, 313-318.

Mertens J.-F. and S. Zamir (1985) Formulation of Bayesian analysis for games with incomplete information, *International Journal of Game Theory*, **14**, 1-29.

Mertens J.-F. and S. Zamir (1995) Incomplete information games and the normal distribution, CRIDT, DP 70.

Mills H. D. (1956) Marginal values of matrix games and linear programs, in *Linear Inequalities and Related Systems*, H. W. Kuhn and A. W. Tucker (eds.), Annals of Mathematical Studies, 38, Princeton University Press, 183-193.

Milnor J. W.and L. S. Shapley (1957) On games of survival, in *Contributions to the Theory of Games, III*, M. Dresher, A.W. Tucker and P. Wolfe (eds.), Annals of Mathematical Studies, 39, Princeton University Press, 15-45.

Monderer D. and S. Sorin (1993) Asymptotic properties in dynamic programming, *International Journal of Game Theory*, **22**, 1-11.

Neyman A. (1998) Nonexpansive mappings and stochastic games, preprint.

Neyman A. and S. Sorin (1997) Equilibria in repeated games of incomplete information: The deterministic symmetric case, in *Game Theoretical Applications to Economics and Operations Research*, T. Parthasarathy and al. (eds.), 129-131, Kluwer Academic Publishers.

Neyman A. and S. Sorin (1998) Equilibria in repeated games of incomplete information: The general symmetric case, *International Journal of Game Theory*, **27**, 201-210.

Nowak A. S. (1985b) Universally measurable strategies in zero-sum stochastic games, *Annals of Probability*, **13**, 269-287.

Nowak A. S. and T.E.S. Raghavan (1991) Positive stochastic games and a theorem of Ornstein, in *Stochastic Games and Related Topics*, T.E. S. Raghavan and al. (eds.), Kluwer, 127-134.

Parthasarathy T. and M. Stern (1976) Markov games - A survey, in *Differential Games and Control Theory, II*, Roxin E.O., P.T. Lin and R. L. Sternberg (eds.), Lecture Notes in Pure and Applied Mathematics, **30**, Marcel Dekker, 1-46.

Parthasarathy T. (1984) Markov games II, *Methods of Operations Research*, **51**, 369-376.

Ponssard J.-P. (1975a) A note on the L-P formulation of zero-sum sequential games with incomplete information, *International Journal of Game Theory*, **4**, 1-5.

Ponssard J.-P. (1975b) Zero-sum games with "almost" perfect information, *Management Science*, **21**, 794-805.

Ponssard J.-P. and S. Sorin (1980a) The L-P formulation of finite zero-sum games with incomplete informatio, *International Journal of Game Theory*, **9**, 99-105.

Ponssard J.-P. and S. Sorin (1980b) Some results on zero-sum games with incomplete information: the dependent case, *International Journal of Game Theory*, **9**, 233-245.

Ponssard J.-P. and S. Sorin (1982) Optimal behavioral strategies in zero-sum games with almost perfect information, *Mathematics of Operations Research*, **7**, 14-31.

Ponssard J.-P. and S. Zamir (1973) Zero-sum sequential games with incomplete information, *International Journal of Game Theory*, **2**, 99-107.

Puterman M. (1994) *Markov Decision Processes*, Wiley.

Rieder U. (1978) On semi-continuous dynamic games, preprint.

Rockafellar R.T. (1970) *Convex Analysis*, Princeton University Press.

Rosenberg D. (1998) Duality and Markovian strategies, *International Journal of Game Theory*, **27**, 577-597.

Rosenberg D. (2000) Zero-sum absorbing games with incomplete information on one side: asymptotic analysis, *SIAM Journal on Control and Optimization*, **39**, 208-225.

Rosenberg D., Solan E. and N. Vieille (2000) Blackwell optimality in Markov decision processes with partial observation, preprint.

Rosenberg D. and S. Sorin (2001) An operator approach to zero-sum repeated games, *Israel Journal of Mathematics*, **121**, 221-246.

Rosenberg D. and N. Vieille (2000) The maxmin of recursive games with lack of information on one side, *Mathematics of Operations Research*, **25**, 23-35.

Scarf, H. and L. S. Shapley (1957) Games with partial information, in *Contributions to the Theory of Games, III*, M. Dresher, A.W. Tucker and P. Wolfe (eds.), Annals of Mathematical Studies, 39, Princeton University Press, 213-229.

Shapley L. S. (1953) Stochastic games, *Proceedings of the National Academy of Sciences of the U.S.A*, **39**, 1095-1100.

Sion M. (1958) On general minimax theorems, *Pacific Journal of Mathematics*, **8**, 171-176.

Sion M. and P. Wolfe (1957) On a game without a value, in *Contibutions to the Theory of Games, III*, M. Dresher, A.W. Tucker and P. Wolfe (eds.), Annals of Mathematical Studies, 39, Princeton University Press, 299-306.

Sorin S. (1979a) An introduction to two-person zero-sum repeated games with incomplete information, *Cahiers du Groupe de Mathématiques Economiques,* **1** (English version, TR 312, IMSS-Economics, Stanford University, 1980).

Sorin S. (1979b) A note on the value of zero-sum sequential repeated games with incomplete information, *International Journal of Game Theory*, **8**, 217-223.

Sorin S. (1984a) Big Match with lack of information on one side (Part I), *International Journal of Game Theory*, **13**, 201-255.

Sorin S. (1984b) On a pair of simultaneous functional equations, *Journal of Mathematical Analysis and Applications*, **98**, 296-303.

Sorin S. (1985a) Big Match with lack of information on one side (Part II), *International Journal of Game Theory*, **14**, 173-204.

Sorin S. (1985b) On a repeated game with state dependent signalling matrices, *International Journal of Game Theory*, **14**, 249-272.

Sorin S. (1989) On repeated games without a recursive structure: existence of $\lim v_n$, *International Journal of Game Theory*, **18**, 45-55.

Sorin S. and S. Zamir (1985) A 2-person game with lack of information on 1 and 1/2 sides, *Mathematics of Operations Research*, **10**, 17-23.

Sorin S. and S. Zamir (1991) Big Match with lack of information on one side, II, in *Stochastic Games and Related Topics*, T.E. S. Raghavan and al. (eds.), Kluwer, 101-112.

Souganidis P.E. (1999) Two player zero sum differential games and viscosity solutions, in *Stochastic and Differential Games*, M. Bardi, T.E.S. Raghavan and T. Parthasarathy eds. Birkhauser, 70-104.

Spinat X. (1999) A necessary and sufficient condition for approachability, *Mathematics of Operations Research*, to appear.

Takahashi M. (1962) Stochastic games with infinitely many strategies, *J. Sci. Hiroshima University Ser. A*, **26**, 123-134.

Vieille N. (1992) Weak approachability, *Mathematics of Operations Research*, **17**, 781-791.

Ville J. (1938) Sur la théorie générale des jeux où intervient l 'habilité des joueurs, in *Traité du Calcul des Probabilités et de ses Applications*, vol. IV, Fascicule II: Applications aux Jeux de Hasard, Borel E. (ed.), Gauthier-Villars, Paris, 105-113.

Von Neumann J. (1928) Zur Theorie der Gesellschaftsspiele, *Mathematische Annalen*, **100**, 295-320.

Waternaux C. (1983a) Solution for a class of repeated games without recursive structure, *International Journal of Game Theory*, **12**, 129-160.

Waternaux C. (1983b) Minmax and maxmin of repeated games without recursive structure, CORE D.P. 8313.

Weyl H. (1950) Elementary proof of a minimax theorem due to von Neumann, in *Contributions to the Theory of Games, I*, H. W. Kuhn and A. W. Tucker (eds.), Annals of Mathematical Studies, 24, Princeton University Press, 19-25.

Yosida K. and S. Kakutani (1940) Markoff process with an enumerable infinite number of possible states, *Japanese Journal of Mathematics*, **XVI**, 47-55.

Zamir S. (1971) On the relation between finitely, and infinitely repeated games with incomplete information, *International Journal of Game Theory*, **1**, 179-198.

Zamir S. (1973a) On repeated games with general information function, *International Journal of Game Theory*, **21**, 215-229.

Zamir S. (1973b) On the notion of value for games with infinitely many stages, *Annals of Statistics*, **1**, 791-796.

Zamir S. (1992) Repeated games of incomplete information : zero-sum, in *Handbook of Game Theory, I*, R.J. Aumann and S. Hart (eds.), North Holland, 109-154.

Yariv L. (1997) A note on repeated games with non monotonic value, *International Journal of Game Theory*, **26**, 229-234.

Index

Déjà parus dans la même collection

1. T. Cazenave, A. Haraux
Introduction aux problèmes d'évolution semi-linéaires. 1990

2. P. Joly
Mise en oeuvre de la méthode des éléments finis. 1990

3/4. E. Godlewski, P.-A. Raviart
Hyperbolic systems of conservation laws. 1991

5/6. Ph. Destuynder
Modélisation mécanique des milieux continus. 1991

7. J. C. Nedelec
Notions sur les techniques d'éléments finis. 1992

8. G. Robin
Algorithmique et cryptographie. 1992

9. D. Lamberton, B. Lapeyre
Introduction au calcul stochastique appliqué. 1992

10. C. Bernardi, Y. Maday
Approximations spectrales de problèmes aux limites elliptiques. 1992

11. V. Genon-Catalot, D. Picard
Eléments de statistique asymptotique. 1993

12. P. Dehornoy
Complexité et décidabilité. 1993

13. O. Kavian
Introduction à la théorie des points critiques. 1994

14. A. Bossavit
Électromagnétisme, en vue de la modélisation. 1994

15. R. Kh. Zeytounian
Modélisation asymptotique en mécanique des fluides newtoniens. 1994

16. D. Bouche, F. Molinet
Méthodes asymptotiques en électromagnétisme. 1994

17. G. Barles
Solutions de viscosité des équations de Hamilton-Jacobi. 1994

18. Q. S. Nguyen
Stabilité des structures élastiques. 1995

19. F. Robert
Les systèmes dynamiques discrets. 1995

20. O. Papini, J. Wolfmann
Algèbre discrète et codes correcteurs. 1995

21. D. Collombier
Plans d'expérience factoriels. 1996

22. G. GAGNEUX, M. MADAUNE-TORT
Analyse mathématique de modèles non linéaires
de l'ingénierie pétrolière. 1996

23. M. DUFLO
Algorithmes stochastiques. 1996

24. P. DESTUYNDER, M. SALAUN
Mathematical Analysis of Thin Plate Models. 1996

25. P. ROUGEE
Mécanique des grandes transformations. 1997

26. L. HÖRMANDER
Lectures on Nonlinear Hyperbolic Differential Equations. 1997

27. J. F. BONNANS, J. C. GILBERT, C. LEMARÉCHAL, C. SAGASTIZÁBAL
Optimisation numérique. 1997

28. C. COCOZZA-THIVENT
Processus stochastiques et fiabilité des systèmes. 1997

29. B. LAPEYRE, É. PARDOUX, R. SENTIS,
Méthodes de Monte-Carlo pour les équations
de transport et de diffusion. 1998

30. P. SAGAUT
Introduction à la simulation des grandes échelles
pour les écoulements de fluide incompressible. 1998

31. E. RIO
Théorie asymptotique des processus aléatoires faiblement dépendants. 1999

32. J. MOREAU, P.-A. DOUDIN, P. CAZES (EDS.)
L'analyse des correspondances et les techniques connexes. 1999

33. B. CHALMOND
Eléments de modélisation pour l'analyse d'images. 1999

34. J. ISTAS
Introduction aux modélisations mathématiques
pour les sciences du vivant. 2000

35. P. ROBERT
Réseaux et files d'attente: méthodes probabilistes. 2000

36. A. ERN, J.-L. GUERMOND
Eléments finis: théorie, applications, mise en œuvre. 2001

Printing: Saladruck, Berlin
Binding: H. Stürtz AG, Würzburg